Applications of Computational Intelligence to Power Systems

Applications of Computational Intelligence to Power Systems

Special Issue Editor

Vassilis S. Kodogiannis

MDPI • Basel • Beijing • Wuhan • Barcelona • Belgrade

MDPI

Special Issue Editor
Vassilis S. Kodogiannis
University of Westminster
UK

Editorial Office
MDPI
St. Alban-Anlage 66
4052 Basel, Switzerland

This is a reprint of articles from the Special Issue published online in the open access journal *Energies* (ISSN 1996-1073) in 2019 (available at: http://www.mdpi.com/journal/energies).

For citation purposes, cite each article independently as indicated on the article page online and as indicated below:

LastName, A.A.; LastName, B.B.; LastName, C.C. Article Title. *Journal Name* **Year**, *Article Number,* Page Range.

ISBN 978-3-03921-760-1 (Pbk)
ISBN 978-3-03921-761-8 (PDF)

Contents

About the Special Issue Editor

Vassilis S. Kodogiannis Reader in Computational Intelligence at the School of Computer Science & Engineering at the University of Westminster, London. He received his Diploma in Electrical Engineering from Democritus University of Thrace, Greece, MSc in VLSI Systems from UMIST, and his PhD in Electrical Engineering from Liverpool University. He was also associated with Technical University of Crete and University of Ioannina, Greece, before joining University of Westminster in 1999. His research interests are in the areas of computational intelligent systems, machine learning, signal and image processing, system identification and control, energy forecasting systems, data mining, robotics, chemometrics, biomedicine, food safety/quality and traffic prediction.

Preface to "Applications of Computational Intelligence to Power Systems"

Nowadays, recent developments and various new challenges in power systems, such as deregulation and environmental concerns, dictate the need for numerous improvements in power systems planning, operation, and control. The requirement of obtaining online solutions as fast as possible for different problems in power systems is becoming more apparent. The electric power industry is continuously searching for ways to improve the efficiency and reliability with which it supplies energy. Although the fundamental technologies of power generation, transmission, and distribution have been changing quite slowly, the power industry has been quick to explore new technologies that might assist its search and to enthusiastically adopting those that show benefits. Traditional methods are usually not able to solve power system problems in real time. In general, such methods are time-consuming and computationally expensive, and are not suitable for online monitoring and control.

Computational intelligence (CI) is one of the most important powerful tools for research in the diverse fields of engineering sciences, ranging from traditional fields of civil and mechanical engineering to vast sections of electrical, electronics, and computer engineering, and above all, the biological and pharmaceutical sciences. The existing field has its origin in the functioning of the human brain in processing information, recognizing patterns, learning from observations and experiments, storing and retrieving information from memory, etc. In particular, as the power industry is on the verge of a changing era, power engineers require computational intelligence tools for the proper planning, operation, and control of power systems. Most of the CI tools are suitably formulated as some sort of optimization or decision-making problem. These CI techniques provide power utilities with innovative solutions for efficient analysis, optimal operation and control, and intelligent decision making.

Due to the nonlinear, interconnected, and complex nature of power system networks and the proliferation of power electronics devices, the CI techniques have become promising candidates for optimal planning, intelligent operation, and automatic control of the power system. Neural networks, deep learning, fuzzy logic, as well as the derivative-free optimization techniques such as genetic algorithm and the swarm intelligence techniques such as particle swarm optimization play an important role in the power industry for decision-making, modeling, and control problems. Due to the nonlinear nature of power system networks and industrial electric systems, fuzzy logic and neural networks are promising candidates for planning, automatic control, system identification, load and load/weather forecasting, etc. Distribution system routing and loss minimization are dealt with effectively using evolutionary algorithms and swarm intelligence techniques. On the other side, the appearance of large generations and highly interconnected systems are making early fault detection and rapid equipment isolation the most important functions for maintaining system stability. One of the factors that hinder the continuous supply of electricity and power is a fault in the power system. These faults cannot be avoided, since they occur because of natural reasons which humans cannot control. CI based methods are being used in the process of fault detection and location to accelerate fault detection and to improve performance of protection system. Developing solutions with these CI tools offers two major advantages: development time is much shorter than when using more traditional approaches, and the systems are very robust, being relatively insensitive to noisy and/or missing data/information known as uncertainty. This special issue deals with different CI techniques for solving real world Power Industry problems. The technical contents will be extremely helpful for the researchers as well as the practicing engineers in the power industry.

Vassilis S. Kodogiannis
Special Issue Editor

energies

MDPI

Article

Integrating Long Short-Term Memory and Genetic Algorithm for Short-Term Load Forecasting

Arpita Samanta Santra [1] and Jun-Lin Lin [1,2,*]

[1] Department of Information Management, Yuan Ze University, Taoyuan 32003, Taiwan; santraarpita83@gmail.com

[2] Innovation Center for Big Data and Digital Convergence, Yuan Ze University, Taoyuan 32003, Taiwan

* Correspondence: jun@saturn.yzu.edu.tw; Tel.: +886-3-463-8800 (ext. 2611)

Received: 18 April 2019; Accepted: 25 May 2019; Published: 28 May 2019

Abstract: Electricity load forecasting is an important task for enhancing energy efficiency and operation reliability of the power system. Forecasting the hourly electricity load of the next day assists in optimizing the resources and minimizing the energy wastage. The main motivation of this study was to improve the robustness of short-term load forecasting (STLF) by utilizing long short-term memory (LSTM) and genetic algorithm (GA). The proposed method is novel: LSTM networks are designed to avoid the problem of long-term dependencies, and GA is used to obtain the optimal LSTM's parameters, which are then applied to predict the hourly electricity load for the next day. The proposed method was trained using actual load and weather data, and the performance results showed that it yielded small mean absolute percentage error on the test data.

Keywords: long short term memory (LSTM); genetic algorithm (GA); short term load forecasting (STLF); electricity load forecasting; multivariate time series

---

1. Introduction

Electricity load forecasting is a mandatory procedure in the capacity planning process of the power industry. Three types of load forecasting are categorized with different time scales. Medium-term and long-term load forecastings predict the weekly, monthly, or yearly electricity load, and both are necessary for system planning, resource investment, and budget allocation. Short-term load forecasting (STLF), the focus of this study, predicts hourly or daily electricity load one hour to one week ahead, which is crucial for resource planning and load balancing in power system management [1–3].

Imprecise load forecasting increases operating cost. For example, an increase of only 1% in forecast error caused an increase of 10 million pounds in operating cost per year for an electric utility company in the United Kingdom [4–6]. Due to the economic and the environmental concerns, electricity load forecasting has drawn considerable attention from both the academic and industrial field. Many approaches have been proposed to solve the STLF problem. For example, many conventional approaches for time series prediction, such as statistical analysis, regression methods [7], smoothing techniques, stochastic process and autoregressive moving average (ARMA) models [8], have been applied to the STLF problem. In last few decades, many data mining approaches have achieved good performance on tackling the uncertainties in STLF, including artificial neural networks (ANN) [9,10], fuzzy inference systems, neuro-fuzzy systems [11], support vector machines, and artificial immune systems [12], etc. Among these data mining approaches, ANN is one of the most popular methods. ANN is suitable for nonlinear problems, has excellent learning ability, and is robust to noise in data.

For time series prediction problems, as in the case of STLF, a special type of ANN called recurrent neural network (RNN) is often used to handle the time-dependency property in time series data. RNN greatly reduces the number of nodes (and consequently, the number of parameters to be learned) in the neural network, but it suffers from the vanishing gradient and long-term dependency problems.

To overcome the drawbacks of the traditional RNN, Hochreiter et al. [13] proposed long short-term memory (LSTM) RNN, which added more controls to the traditional RNN to retain both short-term and long-term memory in the network. Since then, LSTM-based networks have become one of the most promising technologies for deep learning, and have been successfully applied in many research areas such as natural language translation [14], image captioning [15–17], speech recognition [18], and handwriting recognition [19].

With LSTM, the initial weighting of each link in the neural network must be determined before the training iterations start. However, poorly chosen initial weightings could sometimes lead to a bad performance. The motivation of this study was to propose a method such that LSTM could start with a set of properly chosen initial weightings to reduce the forecasting error of the STLF problem. Specifically, the proposed method integrated the searching capability of genetic algorithm (GA) and the learning capability of LSTM. GA is responsible for searching the optimal initial weightings through its population-based bio-inspired evolution operations [20] and LSTM is responsible for learning and memorization intelligence from power usage data and weather data. Overall, the proposed method and the underlying STLF problem can be summarized as follows:

- Input: Weather data such as temperature and humidity are closely related to electricity consumption. Thus, this study used hourly weather data and electricity load for the past 24 h as input in the training process.
- Output: The objective of this study was to produce the hourly electricity load for the next day.
- Methodology framework: The proposed method integrated GA and LSTM. GA was implemented to optimize the parameters of LSTM; LSTM was employed to learn from the past data to yield a better prediction.

The rest of the paper is organized as follows: Section 2 surveys the literature on STLF, and Section 3 describes LSTM for STLF. Section 4 presents our proposed approach. Sections 5 and 6 describe the settings and the results of the performance study, respectively. Section 7 concludes this paper, and gives direction for future research.

2. Related Works on Short-Term Load Forecasting

The problem of load forecasting has attracted much attention since the mid-1960s [21,22]. This section focuses the techniques for STLF. STLF is essentially a time series problem, and thus many traditional time series prediction techniques have been used to solve this problem, e.g., ARMA [23], ARIMA [24], and a hybrid of ARIMA and SVM [25]. Under normal conditions, these statistical techniques deliver good prediction results. However, sudden changes of the weather conditions can dramatically affect the short-term power usage patterns, and consequently, render these statistical techniques failing to provide accurate prediction.

Due to the impact of weather on short-term power usage, many studies have factored into the weather conditions for STLF. Reference [26] applied fuzzy logic [27] to find rules to handle similar weather conditions. Reference [28] used wavelet transform to decompose the electrical load data into components of various frequencies, and then built a SVM model based on the temperature data and the low-frequency components of the electricity load data. Reference [29] built two regression models to respectively predict daily and hourly loads based on the weather data and the electrical load data. Reference [30] applied ANN for long-term load forecasting, and their results showed that ANN is superior to linear regression and SVM.

ANN is a widely used technique for learning nonlinear patterns. Because STLF is a nonlinear problem, many previous works have utilized ANN for STLF, e.g., Reference [31] applied a feed-forward backpropagation ANN for STLF. Hybrids of ANN and other techniques are also common for STLF, e.g., regression tree model [32], time series analysis [33], genetic algorithm [34], chaos genetic algorithm and simulated annealing [35].

With the advance of computing power, deep neural network (DNN) has gained much popularity and been applied to STLF in recent years [36]. LSTM is a special type of DNN that is suitable for time series prediction due to its capability of remembering both the short-term and the long-term behavior in time series data. In Reference [37], two types of LSTM were compared against other deep learning techniques for predicting electricity load of every hour or every minute. Their results showed that LSTM outperformed the other techniques. However, all of the neural network approaches above require setting up the initial weightings of the links in the network, but poorly chosen weightings could lead the searching process trapped in the local optimum. A hybrid of LSTM and GA is proposed in Section 4 to resolve this problem.

3. Long Short-Term Memory for Short-term Load Forecasting

Prior to presenting our approach in Section 4, this section gives a detailed description of LSTM. LSTM is an augmented recurrent neural network model. It learns sequential information with long term dependencies, and preserves information for a long period of time. Traditional recurrent neural network suffers from the vanishing gradient problem. That is, as the number of layers using the same activation function increases, the gradients of the loss function approaches zero, making it difficult to train the network through backpropagation of errors. To prevent the vanishing gradient problem, LSTM utilizes memory cells, where each cell maintains a cell state and a hidden cell state, and uses three gates (namely, input gate, output gate, and forget gate) to control the flow of information into or out of the cell. A formal explanation of the LSTM model is given below.

LSTM is for time series modeling, which maps an input sequence $x = \{x_1, x_2, \ldots, x_n\}$ to an output sequence $y = \{y_1, y_2, \ldots, y_n\}$. For the STLF problem under study, each x_i represents the hourly electricity load and the weather data of day i, and each y_i represents the hourly electricity load of day $i + 1$, indicating a look-ahead parameter of value one. The LSTM contains layers of memory cells, where the interaction between the LSTM layers is shown in Figure 1, and the architecture of a LSTM cell is shown in Figure 2.

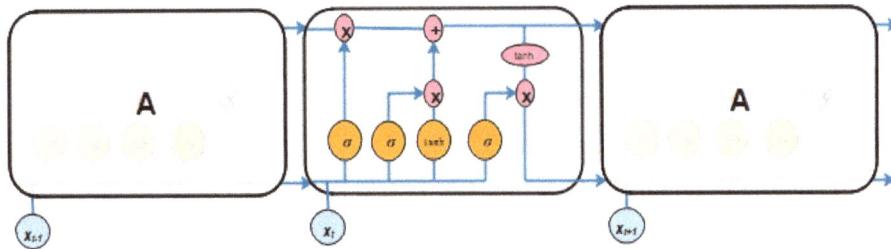

Figure 1. The interaction between Long Short-Term Memory layers.

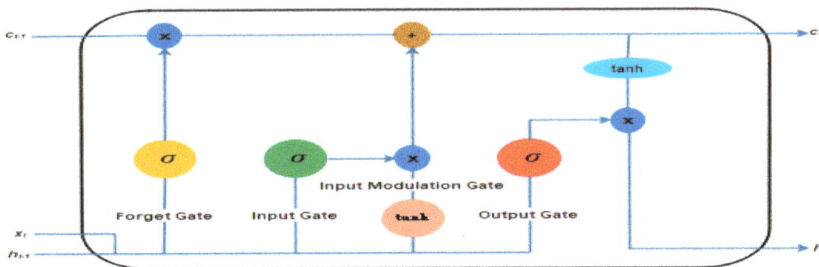

Figure 2. The architecture of the Long Short-Term Memory cell.

3

For ease of exploration, cell t refers to the LSTM cell at the t-th layer in the network. As shown in Figure 2, the input for cell t included x_t, h_{t-1} and c_{t-1}, where x_t was from the input sequence x, and h_{t-1} and c_{t-1} were from the output of cell $t-1$. Directed weighted links connected various components within a LSTM cell or between two adjacent LSTM cells.

The input gate (i_t) of cell t used the Logistic sigmoid function $\sigma(\cdot)$ to decide whether to store the current input x_t and the new cell state in the memory, as shown in Equation (1), where W_{ix}, W_{ih} and W_{ic} were the weights of the incoming links associated with the input gate, and b_i was the bias input of the input gate.

$$i_t = \sigma(W_{ix}x_t + W_{ih}h_{t-1} + W_{ic}c_{t-1} + b_i) \tag{1}$$

Similarly, the forget gate (f_t) of cell t used the Logistic sigmoid function to decide whether to remove the previous cell state (c_{t-1}) from the memory, as shown in Equation (2), where W_{fx}, W_{fh} and W_{fc} were the weights of the incoming links associated with the forget gate, and b_f was the bias input of the forget gate.

$$f_t = \sigma\left(W_{fx}x_t + W_{fh}h_{t-1} + W_{fc}c_{t-1} + b_f\right) \tag{2}$$

The new cell state (c_t) was determined by the amount of old cell state to forget (i.e., $f_t \circ c_{t-1}$) and the amount of new information to include (i.e., $i_t \circ \tanh(W_{cx}x_t + W_{ch}h_{t-1} + b_c)$), as shown in Equation (3), where W_{cx} and W_{ch} were the weights of the incoming links associated with the cell state, b_c was the bias input of the cell state, and $\circ$ represented the hadamard product.

$$c_t = f_t \circ c_{t-1} + i_t \circ \tanh(W_{cx}x_t + W_{ch}h_{t-1} + b_c) \tag{3}$$

The output gate (o_t) used the Logistic sigmoid function to filter information from the current input x_t, previous cell state c_{t-1}, and previous hidden state h_{t-1}, and as shown in Equation (4), where W_{ox}, W_{oh} and W_{oc} were the weights of the incoming links associated with the output gate, and b_o was the bias input of the output gate.

$$o_t = \sigma(W_{ox}x_t + W_{oh}h_{t-1} + W_{oc}c_{t-1} + b_o) \tag{4}$$

The new hidden state (h_t) was calculated as the hadamard product between the output gate values o_t and the tan h function value of the current state c_t, as shown in Equation (5).

$$h_t = o_t \circ \tanh(c_t) \tag{5}$$

Finally, the output of the memory cell (i.e., the predicted value of y_t) was calculated from the hidden state cell state h_t, as shown in Equation (6), where W_{yh} were the weights of the incoming links associated with the hidden state and b_y was the bias input of the hidden state.

$$\hat{y}_t = \left(W_{yh}h_t + b_y\right) \tag{6}$$

For the STLF problem under study, y_t and $\hat{y}_t$ represented the actual and the predicted hourly electricity load of a day, respectively. Thus, both y_t and $\hat{y}_t$ contained 24 values, one for each hour. The predicted error (i.e., $y_t - \hat{y}_t$) was calculated as the mean absolute percentage error (MAPE) of the 24 pairs of corresponding values in y_t and $\hat{y}_t$. The predicted error at time step t was back propagated to refine the weighting matrices in the LSTM. The objective of training the LSTM was to refine the weighting matrices to minimize the predicted error.

Notably, the LSTM contained only one LSTM cell layer, and Figure 1 actually depicts how the cell layer unfolds over time. That is, cell t refers to the status of the LSTM cell at time step t (or day t for the STLF problem under study). Thus, LSTM only maintained a set of weighting matrices and a bias vector for all gates and memory states.

4. Integration of Long Short-Term Memory and Genetic Algorithm: The Proposed Method

As described earlier, the initial values of the weighting matrices could affect the performance of the LSTM. Our proposed method used GA to assist searching the proper initial values for the weighting matrices of the LSTM. GA is a population-based searching technique that employs a population of chromosomes in the searching process. Each chromosome represents a feasible solution. For LSTM, a feasible solution consists of the values of all the weighting matrices described in Section 3. Figure 3 shows the flow diagram of the proposed method. The main steps of the proposed method are described in detail below.

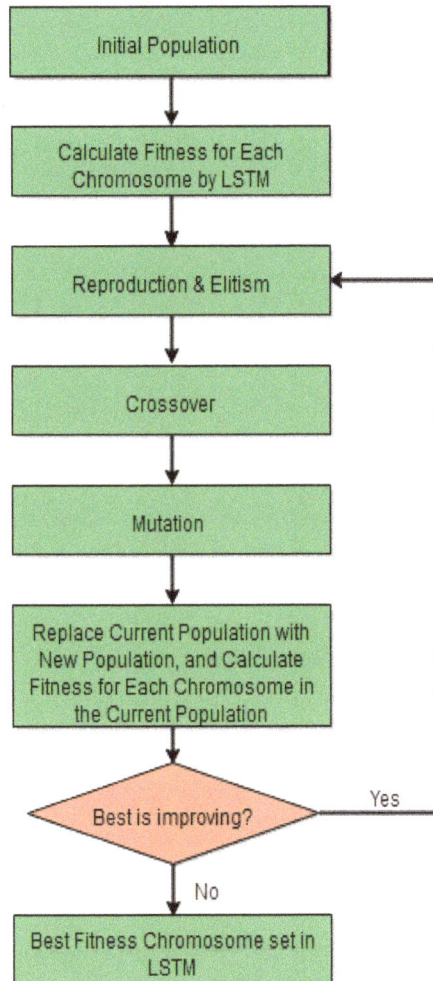

Figure 3. The flow diagram of the proposed method.

Step 1: Generate initial population of chromosomes.

A set of n chromosomes were randomly generated. Each chromosome W contained the values for all the weighting matrices in LSTM, i.e., $W = [W_{ix} \ W_{ih} \ W_{fx} \ W_{fh} \ W_{cx} \ W_{ch} \ W_{ox} \ W_{oh} \ W_{yh}]$.

Step 2: Calculate fitness for each chromosome via LSTM.

For each chromosome in the current population, its value was used to initialize the weighting matrices in the LSTM. Then, training data was fed into the LSTM so that the LSTM could learn from the data to adjust the value of W to minimize the mean absolute percentage error (MAPE) of the training data, as is done in the traditional LSTM technique. The fitness of this chromosome was the final MAPE value of the training data.

Step 3: Generate the new population via genetic operations.

This step generated a new population that contained the same number of chromosomes as the current population. The chromosomes of the new population were generated by applying genetic operations (i.e., reproduction, crossover, or mutation) on the chromosomes selected from the current population. This study used the roulette wheel selection so that a chromosome with a higher fitness value had a higher probability of being selected for genetic operations. In the new population, the proportions of the chromosomes generated via the reproduction, crossover, and mutation operations were referred to as the reproduction, crossover, and mutation ratios, respectively. GA usually adopts a large crossover ratio, and a low mutation ratio.

The reproduction operation repeatedly selected a chromosome from the current population, and added it to the new population, until the reproduction ratio was reached. This study applied an elitism policy so that the best chromosome in the current chromosome was always added to the new population.

The crossover operation repeatedly selected two chromosomes from the current population acting as the parent chromosomes to generate and add two offspring chromosomes to the new population, until the crossover ratio was reached. Uniform crossover was adopted in this study. With uniform crossover, each value in the offspring chromosomes is independently chosen from the two values at the same corresponding position in the two parent chromosomes, as shown in Figure 4.

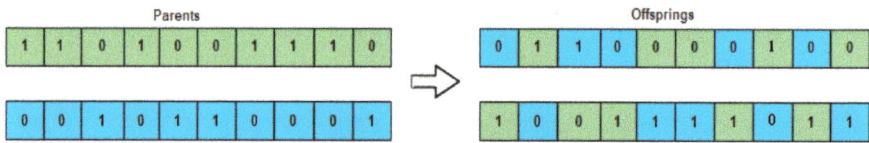

Figure 4. Uniform crossover.

The mutation operation repeatedly selected a chromosome from the current population, modified the selected chromosome to generate a new chromosome (referred to as a mutant), and added the mutant to the new population, until the mutation ratio had reached. One-point mutation was adopted in this study. With one-point mutation, a small random change is injected into the value of a randomly selected position in the selected chromosome to generate the mutant, as shown in Figure 5. The mutation operation introduced diversity to the population of chromosomes

Figure 5. One-point mutation.

Step 4: Replace population and calculate fitness.

At this step, the current population could be abandoned, and the new population became the current population. Similar to Step 2, each chromosome in the current population was used to initialize the weighting matrices in a LSTM, and the training data was fed into the LSTM to yield the fitness value of this chromosome.

Step 5: Stop criterion

If the fitness value of the best chromosome did not improve for *s* continuous generations, then the LSTM of the best chromosome was adopted, and the proposed method was terminated. Otherwise, go to Step 3 to generate a new population. In this study, *s* was set to 50.

5. Experiment Settings

To evaluate the performance of the proposed method, a performance study was conducted on a desktop computer with Intel Core2 Duo E7500@2.94 GHz CPU, 64-bit Windows 10 operating system and 4 GB memory (RAM) using Python 3.6.6. The performance study used hourly electricity load data obtained from the Australian Energy Market Operator, and hourly weather data of the Sydney Observatory obtained from the Bureau of Meteorology (Australia). The timespan of the dataset was from 1 January 2006 to 1 January 2011. The first three years of the dataset was used for training, and the final year was used for testing purposes. The dataset was normalized before being used for training and testing. Because the hourly electricity load patterns were quite different between the weekend and weekday, this study focused only on weekday data.

Each record in the dataset contained the electricity load, dry bulb temperature, dew point temperature, wet bulb temperature, and humidity for every hour within one day. Figure 6 shows that the input of the LSTM network was a record (for day *i*), and the output was the hourly electricity load part of the next record (for day *i* + 1). Thus, the LSTM network had 120 inputs and 24 outputs. The parameter settings of the proposed method were as follows: The number of hidden layers in the LSTM network = 50, the number of epochs for LSTM = 400, population size = 10, crossover ratio = 0.8, and mutation ratio = 0.003.

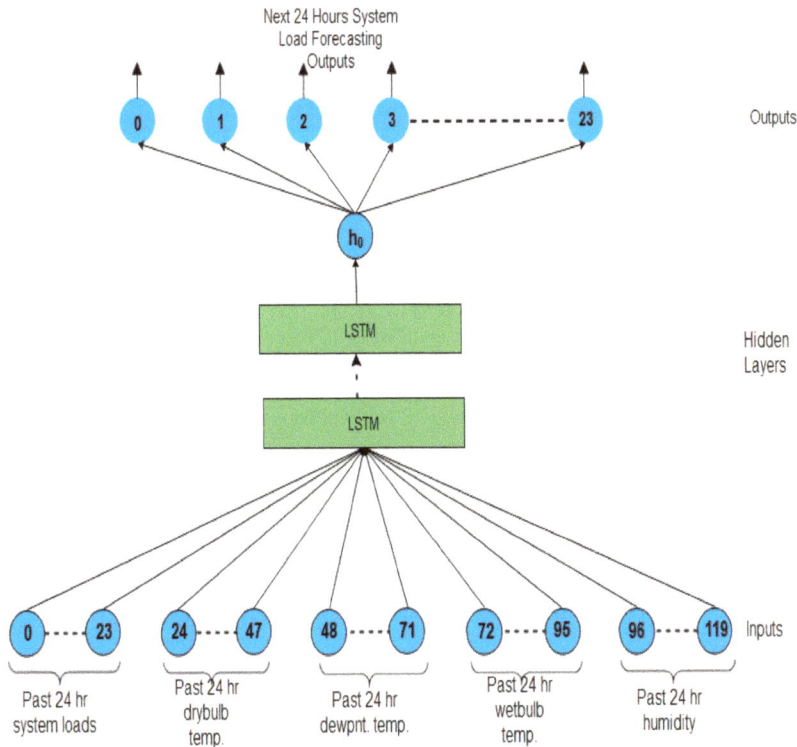

Figure 6. Input and output of the Long Short-Term Memory network for Short-Term Load Forecasting.

6. Experimental Results

In the experimental results, we compared the performance of the proposed method against the LSTM. The forecast results for five randomly selected working days and one randomly selected week in the testing data is plotted in Figure 7. The results showed that the prediction of the proposed methods followed very close with the actual electricity load. LSTM also yielded good results, but it was less stable and could sometimes incur large errors.

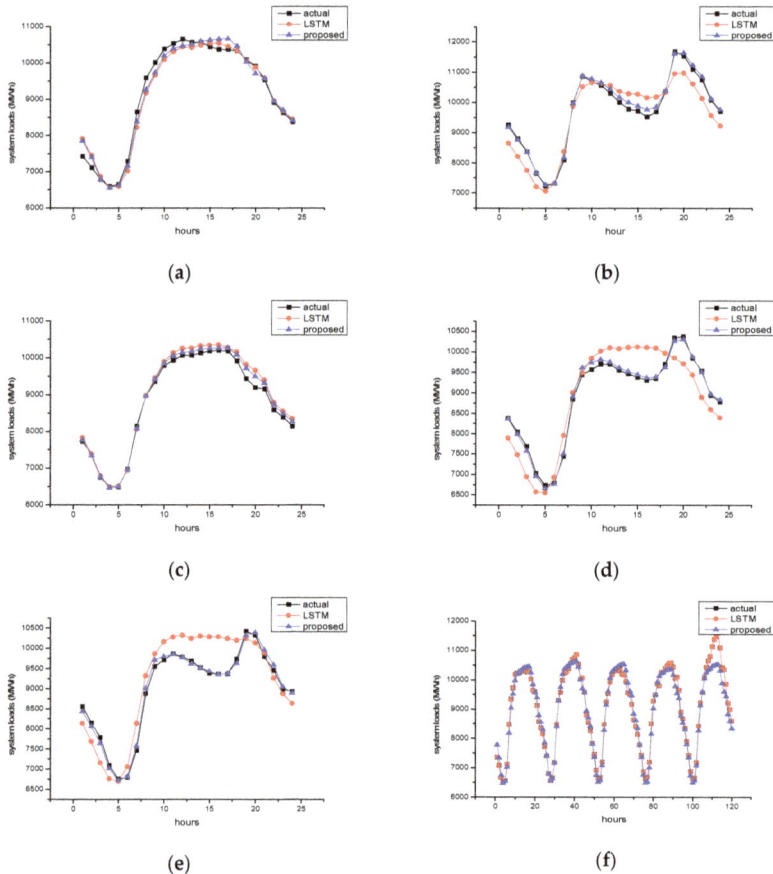

(a)

(b)

(c)

(d)

(e)

(f)

Figure 7. Forecast results for (**a**) 29 March 2010, Monday; (**b**) 8 June 2010, Tuesday; (**c**) 24 November 2010, Wednesday; (**d**) 2 September 2010, Thursday; (**e**) 3 September 2010, Friday; (**f**) the week from 22 March (Monday) to 26 March (Friday) 2010.

Table 1 compares the MAPE of the LSTM and our proposed method for the week of 15 to 19 March 2010. The proposed method consistently performed better than the LSTM and reduced the MAPE of the LSTM by 5.38% to 53.33%. Table 2 compares the MAPE of the LSTM and our proposed method for five randomly chosen days. Similar to Table 1, our proposed method consistently outperformed the LSTM.

The one-year testing data contained 250 records (or weekdays). Figure 8 shows the daily MAPE for the testing data, where the horizontal axis is time, and the vertical axis is MAPE. The results showed that the proposed method consistently improved the LSTM. Descriptive statistics of the 250 daily

MAPEs is shown in Table 3. The proposed method yielded a smaller mean and standard deviation than the LSTM, indicating that the proposed method was more accurate and stable than the LSTM.

Table 1. MAPE of LSTM and proposed method for the week of 15 to 19 March 2010.

Day of the Week	Method		
	LSTM	Proposed Method	Improvement (%)
Mon.	2.94	1.96	33.33
Tue.	1.79	1.19	33.51
Wed.	1.80	0.84	53.33
Thurs.	1.62	1.19	26.54
Fri.	1.30	1.23	5.38

Table 2. MAPE of LSTM and proposed method for five randomly chosen days in the testing data.

Day	Method		
	LSTM	Proposed Method	Improvement (%)
13 Dec. Mon.	2.88	1.84	36.11
8 July. Tue.	4.08	0.87	78.67
17 Feb. Wed.	1.52	1.20	21.05
13 May. Thurs.	3.59	0.86	76.04
27 Aug. Fri.	4.17	1.07	74.34

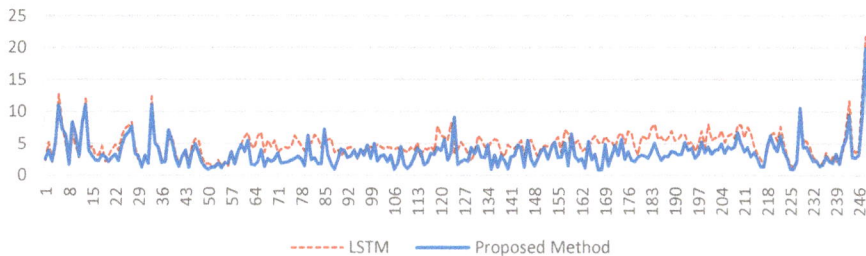

Figure 8. Daily MAPE of the LSTM and the proposed methods for the 250 weekdays in the testing data.

Table 3. Statistics of the daily MAPE for the 250 weekdays in the testing data.

	Method	
	LSTM	Proposed Method
mean	4.8469	3.4889
stdev.	2.2116	2.1506
min	1.2258	0.7879
max	21.7188	19.8419

7. Conclusions

This study proposed a new method that integrated GA and the LSTM for STLF. Although the traditional LSTM has a great learning capability for time series data, it sometimes suffers from the poorly chosen values for its initialization parameters. The proposed method mitigated this problem by applying GA to search suitable values for the initialization parameters of the LSTM. Our performance study showed that the proposed method could effectively improve the prediction accuracy of the LSTM.

This study uses MAPE to measure the error between the predictions and the actual electricity loads. This is based on the assumption that the damage or cost incurred by the error is linearly proportional to the magnitude of the error. However, there are situations that a large prediction error causes a much larger cost or damage to the power companies. For those situations, mean squared error or other more suitable measurements should be adopted, instead of using MAPE.

The electricity demand is easily influenced by many factors. In this study, the weather was used to improve the prediction of the short-term electricity demand. Adding more related factors, e.g., economic conditions, could be of interest for further study. Other meta-heuristic searching techniques and their impact on the execution time are also worthy of investigation for STLF.

Author Contributions: Conceptualization, A.S.S. and J.-L.L.; methodology, J.-L.L.; software, A.S.S.; validation, J.-L.L.; data curation, A.S.S.; visualization, A.S.S.; writing—original draft preparation, A.S.S.; writing—review and editing, J.-L.L.; supervision, J.-L.L.; funding acquisition, J.-L.L. overall contribution: A.S.S. (50%) and J.-L.L. (50%).

Funding: This research was supported by the Ministry of Science and Technology (MOST), Taiwan, under Grant MOST 106-2221-E-155-038.

Conflicts of Interest: The authors declare no conflict of interest.

References

1. Suganthi, L.; Samuel, A.A. Energy models for demand forecasting—A review. *Renew. Sustain. Energy Rev.* **2012**, *16*, 1223–1240. [CrossRef]
2. Al-Hamadi, H.; Soliman, S. Long-term/mid-term electric load forecasting based on short-term correlation and annual growth. *Electr. Power Syst. Res.* **2005**, *74*, 353–361. [CrossRef]
3. Al-Shobaki, S.; Mohsen, M. Modeling and forecasting of electrical power demands for capacity planning. *Energy Convers. Manag.* **2008**, *49*, 3367–3375. [CrossRef]
4. Dudek, G. Artificial immune system for short-term electric load forecasting. In Proceedings of the International Conference on Artificial Intelligence and Soft Computing, Zakopane, Poland, 22–26 June 2008; pp. 1007–1017.
5. Lo, K.; Wu, Y. Risk assessment due to local demand forecast uncertainty in the competitive supply industry. *IEE Proc. Gener. Transm. Distrib.* **2003**, *150*, 573–581. [CrossRef]
6. Soares, L.J.; Medeiros, M.C. Modeling and forecasting short-term electricity load: A comparison of methods with an application to Brazilian data. *Int. J. Forecast.* **2008**, *24*, 630–644. [CrossRef]
7. Yildiz, B.; Bilbao, J.I.; Sproul, A.B. A review and analysis of regression and machine learning models on commercial building electricity load forecasting. *Renew. Sustain. Energy Rev.* **2017**, *73*, 1104–1122. [CrossRef]
8. Pappas, S.S.; Ekonomou, L.; Karampelas, P.; Karamousantas, D.; Katsikas, S.; Chatzarakis, G.; Skafidas, P. Electricity demand load forecasting of the Hellenic power system using an ARMA model. *Electr. Power Syst. Res.* **2010**, *80*, 256–264. [CrossRef]
9. Ekonomou, L.; Oikonomou, D. Application and comparison of several artificial neural networks for forecasting the Hellenic daily electricity demand load. In Proceedings of the 7th WSEAS International Conference on Artificial Intelligence, Knowledge Engineering and Data Bases, Cairo, Egypt, 20–22 February 2008; pp. 67–71.
10. Ekonomou, L.; Christodoulou, C.; Mladenov, V. A short-term load forecasting method using artificial neural networks and wavelet analysis. *Int. J. Power Syst* **2016**, *1*, 64–68.
11. Dudek, G. Neuro-fuzzy approach to the next day load curve forecasting. *Przegląd Elektrotechniczny* **2011**, *87*, 61–64.
12. Dudek, G. Artificial immune system with local feature selection for short-term load forecasting. *IEEE Trans. Evol. Comput.* **2017**, *21*, 116–130. [CrossRef]
13. Hochreiter, S.; Schmidhuber, J. Long short-term memory. *Neural Comput.* **1997**, *9*, 1735–1780. [CrossRef]
14. Sutskever, I.; Vinyals, O.; Le, Q.V. Sequence to sequence learning with neural networks. In Proceedings of the Advances in Neural Information Processing Systems, Montreal, QC, Canada, 8–13 December 2014; pp. 3104–3112.
15. Vinyals, O.; Toshev, A.; Bengio, S.; Erhan, D. Show and tell: A neural image caption generator. In Proceedings of the IEEE Conference on Computer Vision and Pattern Recognition, Boston, MA, USA, 7–12 June 2015; pp. 3156–3164.

16. Karpathy, A.; Fei-Fei, L. Deep visual-semantic alignments for generating image descriptions. In Proceedings of the IEEE Conference on Computer Vision and Pattern Recognition, Boston, MA, USA, 7–12 June 2015; pp. 3128–3137.
17. Mao, J.; Xu, W.; Yang, Y.; Wang, J.; Huang, Z.; Yuille, A. Deep captioning with multimodal recurrent neural networks (m-rnn). *arXiv* **2014**, arXiv:1412.6632.
18. Graves, A.; Jaitly, N. Towards end-to-end speech recognition with recurrent neural networks. In Proceedings of the International Conference on Machine Learning, Beijing, China, 21–26 June 2013; pp. 1764–1772.
19. Graves, A.; Schmidhuber, J. Offline handwriting recognition with multidimensional recurrent neural networks. In Proceedings of the Advances in Neural Information Processing Systems, Vancouver, BC, Canada, 8–10 December 2009; pp. 545–552.
20. Mitchell, M. *An Introduction to Genetic Algorithms*; MIT Press: Cambridge, MA, USA, 1998.
21. Khuntia, S.R.; Rueda, J.L.; van der Meijden, M.A. Forecasting the load of electrical power systems in mid-and long-term horizons: A review. *IET Gener. Transm. Distrib.* **2016**, *10*, 3971–3977. [CrossRef]
22. Hahn, H.; Meyer-Nieberg, S.; Pickl, S. Electric load forecasting methods: Tools for decision making. *Eur. J. Oper. Res.* **2009**, *199*, 902–907. [CrossRef]
23. Huang, S.-J.; Shih, K.-R. Short-term load forecasting via ARMA model identification including non-Gaussian process considerations. *IEEE Trans. Power Syst.* **2003**, *18*, 673–679. [CrossRef]
24. Pappas, S.S.; Ekonomou, L.; Moussas, V.; Karampelas, P.; Katsikas, S. Adaptive load forecasting of the Hellenic electric grid. *J. Zhejiang Univ. Sci. A* **2008**, *9*, 1724–1730. [CrossRef]
25. Karthika, S.; Margaret, V.; Balaraman, K. Hybrid short term load forecasting using ARIMA-SVM. In Proceedings of the 2017 Innovations in Power and Advanced Computing Technologies (i-PACT), Vellore, India, 21–22 April 2017; pp. 1–7.
26. Mukhopadhyay, P.; Mitra, G.; Banerjee, S.; Mukherjee, G. Electricity load forecasting using fuzzy logic: Short term load forecasting factoring weather parameter. In Proceedings of the 2017 7th International Conference on Power Systems (ICPS), Pune, India, 21–23 December 2017; pp. 812–819.
27. Roman, R.-C.; Precup, R.-E.; David, R.-C. Second order intelligent proportional-integral fuzzy control of twin rotor aerodynamic systems. *Procedia Comput. Sci.* **2018**, *139*, 372–380. [CrossRef]
28. Li, Y.; Fang, T. Wavelet and support vector machines for short-term electrical load forecasting. In *Wavelet Analysis and Its Applications: (In 2 Volumes)*; World Scientific: Singapore, 2003; pp. 399–404.
29. Ruzic, S.; Vuckovic, A.; Nikolic, N. Weather sensitive method for short term load forecasting in electric power utility of Serbia. *IEEE Trans. Power Syst.* **2003**, *18*, 1581–1586. [CrossRef]
30. Ekonomou, L. Greek long-term energy consumption prediction using artificial neural networks. *Energy* **2010**, *35*, 512–517. [CrossRef]
31. McClelland, J.L.; Rumelhart, D.E.; Group, P.R. Parallel distributed processing. *Explor. Microstruct. Cogn.* **1986**, *2*, 216–271.
32. Yang, J.; Stenzel, J. Short-term load forecasting with increment regression tree. *Electr. Power Syst. Res.* **2006**, *76*, 880–888. [CrossRef]
33. Deng, J.; Jirutitijaroen, P. Short-term load forecasting using time series analysis: A case study for Singapore. In Proceedings of the 2010 IEEE Conference on Cybernetics and Intelligent Systems, Singapore, 28–30 June 2010; pp. 231–236.
34. Ling, S.-H.; Leung, F.H.-F.; Lam, H.-K.; Lee, Y.-S.; Tam, P.K.-S. A novel genetic-algorithm-based neural network for short-term load forecasting. *IEEE Trans. Ind. Electron.* **2003**, *50*, 793–799. [CrossRef]
35. Liao, G.-C.; Tsao, T.-P. Application of a fuzzy neural network combined with a chaos genetic algorithm and simulated annealing to short-term load forecasting. *IEEE Trans. Evol. Comput.* **2006**, *10*, 330–340. [CrossRef]
36. Ryu, S.; Noh, J.; Kim, H. Deep neural network based demand side short term load forecasting. *Energies* **2016**, *10*, 3. [CrossRef]
37. Marino, D.L.; Amarasinghe, K.; Manic, M. Building energy load forecasting using deep neural networks. In Proceedings of the IECON 2016-42nd Annual Conference of the IEEE Industrial Electronics Society, Florence, Italy, 23–26 October 2016; pp. 7046–7051.

energies

MDPI

Article

An Adaptive Particle Swarm Optimization Algorithm Based on Guiding Strategy and Its Application in Reactive Power Optimization

Fengli Jiang [1], Yichi Zhang [2], Yu Zhang [3], Xiaomeng Liu [1] and Chunling Chen [1,*]

[1] College of Information and Electrical Engineering, Shenyang Agricultural University, Shenyang 110866, China; fengli0308@163.com (F.J.); LXM_SN@163.com (X.L.)
[2] School of Electrical Engineering, Southwest Jiaotong University, Chengdu 610031, China; yichizhang@my.swjtu.edu.cn
[3] Anhui Provincial Laboratory of New Energy Utilization and Energy Conservation, Hefei University Technology, Hefei 230009, China; zhangyu1994@mail.hfut.edu.cn
* Correspondence: snccl@163.com; Tel.: +86-155-6603-0403

Received: 2 April 2019; Accepted: 29 April 2019; Published: 5 May 2019

Abstract: An improved adaptive particle swarm algorithm with guiding strategy (GSAPSO) was proposed, and it was applied to solve the reactive power optimization (RPO). Four kinds of particles containing the main particles, double central particles, cooperative particles and chaos particles were introduced into the population of the developed algorithm, which was to decrease the randomness and promote search efficiency through guiding particle position updating. Moreover, the cluster focus distance-changing rate was responsible for dynamically adjusting inertia weight. Then the convergence rate and accuracy of this algorithm would be elevated by four functions, which would test effectively the proposed. Finally, the optimized algorithm was verified on the RPO of the IEEE 30-bus power system. The performance of PSO, Random weight particle swarm optimization (WPSO) and Linearly decreasing weight of the particle swarm optimization algorithm (LDWPSO) were identified as the referential information, the proposed GSAPSO was more efficient from the comparison. Calculation results demonstrated that higher quality solutions were obtained and convergence rate and accuracy was significantly higher with regard to the GSAPSO algorithm.

Keywords: particle swarm optimization; particle update mode; inertia weight; reactive power optimization

1. Introduction

Reactive power optimization (RPO) has become more and more crucial as a segment of optimal power flow calculation, with the rapid development of China's power grid and the continuous expansion of its scale. The RPO is equal to reasonably distributing reactive power generation for minimizing the active power losses and maintaining all bus voltages within a reasonable range, while satisfying some constraints [1–4]. The way to achieve the aforementioned aims depends on the capacity of shunt capacitors, the output of generators and the adjustment of transformer tap positions [5–8].

Generally, the nonlinear characteristic is prominent in RPO and is composed of multivariate, multi-constraint, discrete variables and continuous variables. A number of conventional optimization methods or techniques, traditional linear programming (LP) [9], Newton method [10], quadratic programming (QP) [11], and interior point method (IPM) [12], were put forward to deal with RPO. However, due to discontinuity and multi-peak, these techniques did not handle well. In recent years, some of the heuristic algorithms such as genetic algorithm (GA) [8,13], simulated annealing (SA) [14], differential evolution (DE) [15,16], improved shuffled frog leaping algorithm (ISFLA) [17], gray wolf optimizer (GWO) [18], particle swarm optimization (PSO) [19–22] have been widely developed for solving RPO, and have achieved good results. As a clustering intelligence algorithm, the PSO algorithm simulates the behaviour of bird flock foraging, and achieves the goal of optimization through the cooperation and competition among flock birds [23]. Compared to other heuristic techniques, the PSO has a few adjustment parameters and a simple implementation. However, with the increase of variables and more local optimum in the objective function, the PSO algorithm is susceptible to prematurely converging and being put into local optimum. Many researchers have proposed and published potential improvements to the PSO algorithm, which can improve the global and local search ability and avoid the above phenomenon. Kannan et al. [24] proposed a solution for the capacitor optimization configuration in radial distribution using Multi-Agent Particle Swarm Optimization. Zad et al. [25] considered the reactive power control of distributed generations by implementing PSO. Achayuthakan et al. [26] proposed a PSO for RPO achieving reactive power cost allocation. Esmin et al. [22] presented a method to lessen power loss via a hybrid PSO with a mutation operator. Differentially perturbed velocity was used to improve PSO for the ORPD problem of a power system in [27]. Singh et al. [19] presented a new PSO for the scheme of optimal reactive power dispatch (ORPD), where improvement was the introduction of an aging leader and challengers.

In this paper, an improved PSO based on a guiding strategy ws developed, which optimized the particle update mode and the adjustment mechanism of inertia weight, which had no negligible impact on the PSO algorithm. In addition, it was applied in order to solve the PRO in a power system and the standard IEEE 30-bus power system was employed. The effectiveness of the GSAPSO was contrasted to classical PSO, random weight PSO (WPSO) and linearly decreasing weight PSO (LDWPSO).

The structure of this paper is presented as follows: A brief description of PSO is brought out in Section 2. Section 3 presents details of improved PSO. Section 4 provides referential methods to test the three PSO algorithms and the comparison is discussed. Section 5 introduces the formulation of the RPO problem, and the solution algorithms are outlined in Section 6. Section 7 exhibits the simulation results showing the solution effectiveness. Conclusions are given in Section 8.

2. Original Particle Swarm Optimization (PSO)

PSO defines the variables in the solution space as particles without weight and volume. Each particle has two important elements: Velocity and position. The velocity makes a dynamic adjustment on the basis of the flight experience of individuals and groups. The position refers to a potential solution, and the corresponding fitness value can be obtained to judge the quality of the position. Each particle determines the velocity and direction of the flight according to its current location, the optimal location of the individual and the group.

In the S-dimensional space, a randomly initialized population with m particles is generated. The information for each particle i contains its position $(X_i = [x_{i1}, x_{i2}, \ldots, x_{is}])$ and velocity $(V_i = [v_{i1}, v_{i2}, \ldots, v_{is}])$, the best value of position for particle i is referred to as $P_i = [p_{i1}, p_{i2}, \ldots, p_{is}]$, whose fitness value is expressed as 'pbest'. Moreover, the best value of the position for the group is recorded as $P_g = [p_{g1}, p_{g2}, \ldots, p_{gs}]$, and its fitness value is expressed as 'gbest'. In the searching procedure, the velocity and position of each particle are constantly updated through Equations (1) and (2) respectively.

$$V_{id}^{k+1} = \omega^k V_{id}^k + c_1 r_1 (P_{id}^k - X_{id}^k) + c_2 r_2 (P_{gd}^k - X_{id}^k) \tag{1}$$

$$X_{id}^{k+1} = X_{id}^k + V_{id}^{k+1} \tag{2}$$

where k is time of current iteration; r_1 and r_2 are random variables taken between 0 and 1; ω^k is the inertia weight at the kth iteration. c_1 and c_2 are acceleration coefficients. P_{id}^k is the best value of the ith particle in d dimension. P_{gd}^k is the best value among the entire population. The velocity range of the particles should be set in case the particle flies out of the search space too quickly and improve optimization ability simultaneously.

3. Adaptive Particle Swarm Algorithm with Guiding Strategy(GSAPSO)

In order to prevent blind search during the early stage and slow search speed as well as easily trap in the local optimum in the later period, the GSAPSO was proposed. The GSAPSO was mainly improved in two aspects, one was the mixed particles update mode, on the other hand, the inertia weight was improved.

3.1. Strategy for Mixed Particles Update

The standard PSO algorithm initializes the population randomly and has strong randomness. To get better performance in convergence, the particles were divided into four parts in this paper. The first part was the main particles randomly generated in system. The second part consisted of two kinds of central particles [28], which were the central particles produced by the first part and the central particles formed by the individual optimal value of all the particles, respectively.

Usually, the optimal solution has a higher probability of appearing in the center position. Therefore, the central particles are used to accelerate the convergence. In the whole search process, the properties of them were the same as those of ordinary particles, except that they had no velocity characteristics.

In the searching process, the two particles with the worst fitness value in the main particles were replaced by the double central particles, and then the particles were prone to fly towards better searching areas.

The third part was cooperative particles. The idea was to make multiple independent particles start from the global optimal position, first updated in one dimension, and then extra particles were updated randomly in the D dimension. The cooperative particles were used to enhance the capability in global search. The fourth part introduced chaotic particles, generated by logistic mapping. The chaotic particles were used to search for sub-optimal solutions of particles in the region around the boundary. The idea of updating the mixed particles was to exchange the good values obtained by the second, third, and fourth parts with the poor values of the first part during each iteration, so that the whole particles approached the best solution.

(1) The central particles were expressed as Equations (3) and (4).

$$x_{SCP} = 1/n \sum_{i=1}^{n} x_{id} \tag{3}$$

$$x_{GCP} = 1/n \sum_{i=1}^{n} P_{id} \tag{4}$$

where n represents the quantity of main particles; x_{SCP} *and* x_{GCP} are the central particle produced by the main particles and the individual optimal values, respectively.
(2) The cooperative particles update formula was as follows

$$x_{id} = P_{gd} + rand \times vec \tag{5}$$

$$vec = V_{imax} \times (1 - \alpha \times \frac{exetime}{gen}) \tag{6}$$

where *vec* is a velocity variable whose value can be obtained according to Equation (6). The *exetime*, *gen* represents the current number and the total number of iterations respectively. α is the update coefficient. The *rand* is a random variable taken between 0 and 1 and obeys uniform distribution.
(3) The chaotic particles were determined as follows

$$\begin{cases} k = 4 \times k \times (1 - k) \\ x_{id} = x_{imin} + k \times (x_{imax} - x_{imin}) \end{cases} \tag{7}$$

where x_{imax}, x_{imin} are the maximum and minimum of the coordinates. k is the chaotic factor.

The update of the first part of the main particles was obtained by Equations (1) and (2); the update of the second part of central particles could be obtained by the Equations (3) and (4); and the third part of cooperative particles update formula was shown by Equations (5) and (6). The fourth part of chaotic particles could be updated by Equation (7).

3.2. Strategy for Inertia Weight

ω^k controls the influence of speed in the kth iteration on the speed in the $k+1$th iteration, and ω of the standard PSO algorithm can be regarded as 1.0. The larger the inertia term value, the stronger the global search ability. Conversely, the smaller the inertia term value, the stronger the local search ability. Therefore, the size of ω could be dynamically adjusted according to different periods during the iterative process to balance the ability of local and global explorations. In [29], the LDWPSO was proposed, which linearly reduced the ω value during the iterative process. The principle of the LDWPSO algorithm was that ω was larger for the global search in the initial iteration, while the ω value in the later iteration was smaller, and it was easily put into the local optimum. In [30], a dynamic adaptive nonlinear inertia weight was introduced into the PSO. The algorithm judged the particle distribution of the population by the size of the focus distance-changing rate, and then updated dynamically the nonlinear inertia weight according to different situations, which updated the fitness of particles at a suitable speed, and had the capability of good global search ensuring the convergence speed.

The adjustment strategy of inertia weight was to find the Euclidean spatial distance between all particles and the historical optimal in the population. The distance represented the closeness of the particles to the optimal position. The formulas of the average and maximum focus distance are described as follows:

$$Dist_{mean} = \frac{\sum\limits_{i=1}^{m} \sqrt{\sum\limits_{d=1}^{D} (X_{id} - P_{gd})^2}}{m} \tag{8}$$

$$Dist_{max} = \max_{i=1,2,...n} \left(\sqrt{\sum\limits_{d=1}^{D} (X_{id} - P_{gd})^2} \right) \tag{9}$$

where m is the quantity of particles in the population. D is the quantity of dimensions of each particle. P_{gd} is the best position of the population. X_{id} is the best position of the individual.

Then we could get the current focus distance-changing rate of the particle:

$$k = \frac{Dist_{max} - Dist_{mean}}{Dist_{max}} \tag{10}$$

In the iterative process, the inertia weight was adjusted by the focus distance changing rate, thereby adjusting the ability of the global and the local search. The inertia weight update in [30] was not ideal. After continuous experiment, it was found that the convergence of the algorithm was excellent when the variation curve of the inertia weight with k showed a downward convex trend. Therefore, the Formula (11) was adapted to calculate ω.

$$\omega = \begin{cases} \ln(a - k) + b & k \geq 0.05 \\ 0.75 + rand/4 & k < 0.05 \end{cases} \tag{11}$$

where the *rand* was a random variable taken between 0 and 1 and obeyed uniform distribution. Compared with the linear decreasing weight method, this selection strategy could better adapt to the complex actual environment, adjust the ability of global and local search more flexibly, and the diversity of the population could be preserved simultaneously.

4. Experiment

4.1. Experimental Setup

In this section, the comparison of optimized algorithm and WPSO and LDWPSO was carried out by means of classical Benchmark functions consisting of the Sphere function, Rosenbrock function, Rastrigin function and Griewank function, which the performance test generally used. The details of four functions are presented as Figure 1:

(1) Sphere Function

The Sphere function is nonlinear and symmetric, it has a single-peak, which can be separated with different dimensions. It is used to test the optimization precision.

(2) Rosenbrock Function

The Rosenbrock function is a typical ill-conditioned function that is difficult to minimize. Since this function provides little information for the search, it is difficult to identify the search direction, and the chance of finding the global optimal solution is low. Therefore, it is often used to evaluate the execution performance.

(3) Rastrigin Function

The Rastrigin function is a complex multi-peaks function, with large quantity of local optima. It is prone to making the algorithm put local optimum, which will prematurely converge and not get the global optimal solution.

(4) Griewank Function

The Griewank function is also a complex multi-peaks function with a large quantity of local minimum.

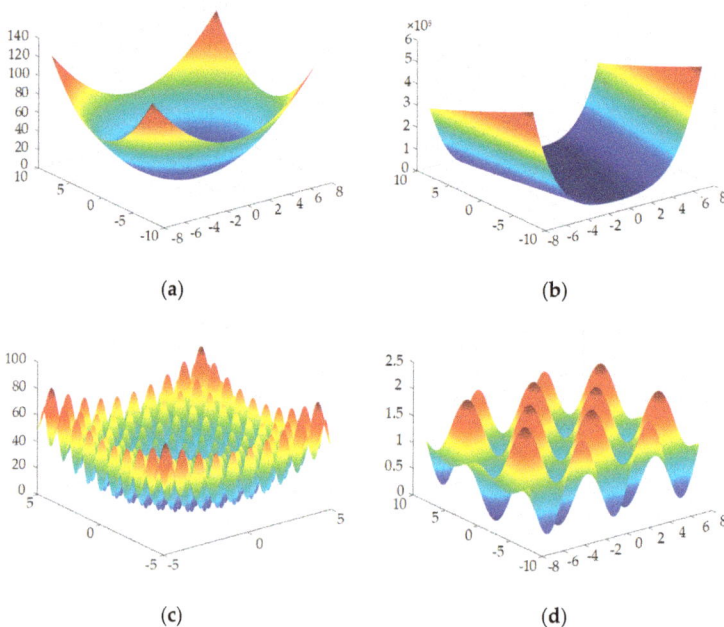

(a)

(b)

(c)

(d)

Figure 1. Three-dimensional map of the four functions: (**a**) Sphere function; (**b**) Rosenbrock function; (**c**) Rastrigin function; (**d**) Griewank function.

4.2. Parameter Settings

For the four functions, the optimal objective values were all zero. The maximum number of iterations in the first and fourth functions was set to 1000, and that was 2000 in the second and third functions. The population size was 60, the main particle accounted for 70%, including two central particles, the cooperative particles accounted for 20%, and the chaotic particles accounted for 10%. Learning factor $c_1 = c_2 = 2$, inertia weight in LDWPSO algorithm was $\omega_{max} = 0.9$, $\omega_{min} = 0.3$. Each function used each algorithm to run 10 times independently. Table 1 shows the relevant information of four functions.

Table 1. Related information of four functions.

Function	Function Expression	Dimension	Range
Sphere	$f_1 = \sum_{i=1}^{n} x_i^2$	30	$[-100, 100]^n$
Rosenbrock	$f_2 = \sum_{i=1}^{n-1}\left[100(x_{i+1} - x_i^2)^2 + (x_i - 1)^2\right]$	10	$[-30, 30]^n$
Rastrigin	$f_3 = \sum_{i=1}^{n}\left[x_i^2 - 10\cos(2\pi x_i) + 10\right]$	10	$[-5.12, 5.12]^n$
Griewank	$f_4 = \frac{1}{4000}\sum_{i=1}^{n} x_i^2 - \prod_{i=1}^{n}\cos(\frac{x_i}{\sqrt{i}}) + 1$	30	$[-600, 600]^n$

4.3. Analysis of Test Results

For LDWPSO, WPSO and the GSAPSO algorithms, the convergence accuracy and speed of the algorithms were analyzed respectively. The statistical results of the experiment are shown in Tables 2 and 3.

Table 2. Convergence accuracy of GSAPSO, LDWPSO, WPSO algorithms.

Functions	GSAPSO			LDWPSO			WPSO		
	BEST	WORST	MEAN	BEST	WORST	MEAN	BEST	WORST	MEAN
Sphere	9.38×10^{-27}	2.81×10^{-22}	4.86×10^{-23}	7.27×10^{-13}	1.85×10^{-11}	9.53×10^{-12}	1.01×10^{-4}	1.36×10^{-2}	2.73×10^{-3}
Rosenbrock	1.25×10^{-4}	2.29	1.31	0.0837	5.795	2.975	1.81	10.187	4
Rastrigin	0	0	0	0	2.985	1.393	0	3.98	1.89
Griewank	0	0.0465	9.33×10^{-3}	7.7×10^{-13}	0.032	0.0133	6.49×10^{-4}	0.0623	0.0222

Table 3. Convergence speed of GSAPSO, LDWPSO and WPSO algorithms.

Functions	GSAPSO			LDWPSO			WPSO		
	AT	BT	SR	AT	BT	SR	AT	BT	SR
Sphere	148	119	100%	644	623	100%	685	571	90%
Rosenbrock	1626	118	20%	2000	1090	-	2000	1944	-
Rastrigin	126	84	100%	1762	1081	20%	1704	513	20%
Griewank	321	125	80%	790	622	60%	871	620	40%

(1) Algorithm convergence accuracy analysis

According to the parameter setting in Section 4.2, the convergence precision of the three PSO algorithms was counted in the same conditions. The best value (BEST), the mean value (MEAN), and the worst value (WORST) of the objective function in ten iterations could characterize the performance of each algorithm. From Table 2, it demonstrated that the GSAPSO was superior to the others for the BEST, the MEAN, and the WORST. Moreover, the three types of values of GSAPSO algorithm were not much different, which indicated that the GSAPSO had better stability. The results of the Sphere function reflected the GSAPSO algorithm was much faster than the LDWPSO and WPSO algorithms to obtain the optimal value. The Rosenbrock function was more complex, and the optimization result not only reflected the convergence speed of the algorithm, but reflected the ability of jumping out local optimum. Therefore, it could be seen that the GSAPSO had a higher convergence accuracy than the others. For Rastrigin and Griewank functions, the GSAPSO could accurately find the optimal value, while the other two algorithms could not get the optimal value for the Griewank function.

(2) Analysis of algorithm convergence speed

According to the parameter setting in Section 4.2, the convergence speed of the three algorithms was counted in the same conditions. The average of iterations (AT), the minimum iterations (BT) and the ratio of the number for satisfying the convergence criteria to the total number of experiments (SR) were calculated when each algorithm converged to 10^{-2} in 10 iterations. Table 3 shows that the values of AT and BT of the GSAPSO significantly decreased compared to the LDWPSO and WPSO, and the average time required was less for the GSAPSO. It demonstrated that the GSAPSO had a higher computational efficiency. In short, the GSAPSO had superiority in the success rate and computational efficiency compared with the other two algorithms.

The average adaptation curves of the four standard test functions using the three PSO algorithms are shown in Figure 2.

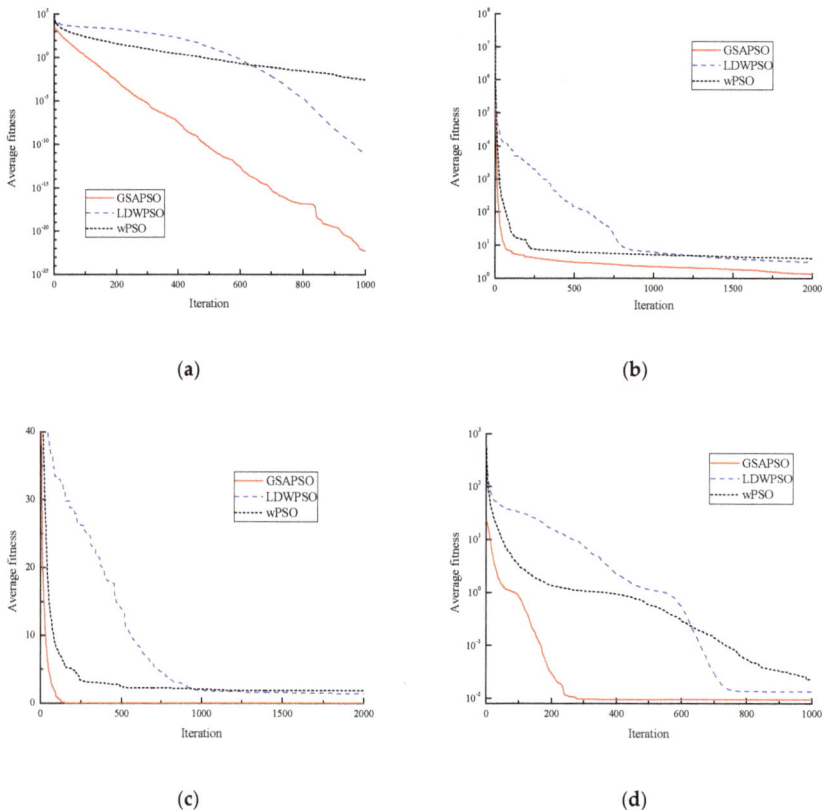

(a)

(b)

(c)

(d)

Figure 2. Function fitness curve: (**a**) Sphere fitness curve; (**b**) Rosenbrock fitness curve; (**c**) Rastrigin fitness curve; (**d**) Griewank fitness curve.

From Figure 2, it shows that the GSAPSO algorithm had greatly improved on convergence accuracy and convergence time compared to the LDWPSO and WPSO algorithms, which increased the chance of finding the optimal value for the actual optimization problem.

5. Problem Formulations

5.1. Objective Functions

Minimizing the active power losses and maintaining voltage stability are the purposes of the RPO to power system. The bus voltage and the reactive power of the generators both had a certain range of limits, which also would have had an impact on the results at the same time, so the two state variables were added to the objective function as quadratic penalty terms. The RPO problem was formulated as:

$$\min f = \min P_{loss} = \sum_{i,j \in N_B} g_{ij}(U_i^2 + U_j^2 - 2U_iU_j\cos\theta_{ij}) + \lambda_1 \sum_{i=1}^{N_{PQ}} \left(\frac{U_i - U_{i\lim}}{U_{i\max} - U_{i\min}}\right)^2 + \lambda_2 \sum_{i=1}^{N_{PV}} \left(\frac{Q_{Gi} - Q_{Gi,\lim}}{Q_{Gi,\max} - Q_{Gi,\min}}\right)^2 \quad (12)$$

where U_i is the voltage magnitude at bus i. N_B, N_{PQ}, N_{PV} are the total of buses, the set of PQ buses and the set of PV buses respectively. λ_1, λ_2 are the penalty terms, which are assigned according to respective importance of variables. In this paper, it was assumed $\lambda_1 = 100$, $\lambda_2 = 10$. g_{ij}, θ_{ij} were the mutual conductance and voltage angle difference between bus i and j. $U_{i\lim}$, $Q_{Gi,\lim}$ in Equation (12) was defined as follows:

$$U_{i\lim} = \begin{cases} U_{i\max}, & U_i > U_{i\max} \\ U_{i\min}, & U_i < U_{i\min} \end{cases} \quad (13)$$

$$Q_{Gi,\lim} = \begin{cases} Q_{Gi,\max}, & Q_{Gi} > Q_{Gi,\max} \\ Q_{Gi,\min}, & Q_{Gi} < Q_{Gi,\min} \end{cases} \quad (14)$$

5.2. Constraints

The constraints were segmented into two parts, which are described as follows:
Equality constraints:

$$\begin{cases} P_i = U_i \sum_{j=1}^{N_B} U_j(G_{ij}\cos\theta_{ij} + B_{ij}\sin\theta_{ij}) & i \in N_{PV}, N_{PQ} \\ Q_i = U_i \sum_{j=1}^{N_B} U_j(G_{ij}\sin\theta_{ij} - B_{ij}\cos\theta_{ij}) & i \in N_{PQ} \end{cases} \quad (15)$$

Inequality constraints (control variables):

$$\begin{cases} U_{Gi,\min} \leq U_{Gi} \leq U_{Gi,\max}, & i \in N_G \\ Q_{Ci,\min} \leq Q_{Ci} \leq Q_{Ci,\max}, & i \in N_C \\ T_{k,\min} \leq T_k \leq T_{k,\max}, & k \in N_T \end{cases} \quad (16)$$

where N_G, N_C, N_T are the quantity of generators, capacitor banks and transformers respectively. U_{Gi} is the voltage magnitude at generator bus i. Q_{Ci} is the compensation capacity of ith shunt capacitor; T_k is the tap position of kth transformer.

The inequality constraints on state variables were as follows:

$$\begin{cases} U_{i,\min} \leq U_i \leq U_{i,\max} & i \in N_{PQ} \\ Q_{Gi,\min} \leq Q_{Gi} \leq Q_{Gi,\max} & i \in N_G \end{cases} \quad (17)$$

where Q_{Gi} is the reactive power of ith generator; $U_{i\min}$, $U_{i\max}$, $Q_{Gi,\min}$, $Q_{Gi,\max}$ are the minimum and maximum of the corresponding variables respectively.

6. GSAPSO Algorithm for RPO

6.1. Treatment of Control Variables

The voltage of the generator is a continuous variable, the transformer tap position and the capacity of capacitor banks are discrete variables. Continuous variables were encoded in real numbers, and integers were applied to encode discrete variables in this paper.

$$X = [U_G; T; C] = [U_{G1}, \cdots, U_{Gi}, \cdots U_{GN}; T_1, \cdots, T_i, \cdots T_M; C_1, \cdots, C_i, \cdots C_K] \tag{18}$$

where X is individual position of particles; U_{Gi} is the voltage magnitude of the ith generator. T_i is the tap ratio of ith transformer. C_i is the capacity of ith capacitor banks. The subscripts N, M, and K represent the quantity of generator buses, adjustable transformers, and compensation buses, respectively.

6.2. Treatment of Discrete Variables

This study uses the rounding method to process discrete variables. Their formulas are:

$$\begin{aligned} T_i &= round(rand \times stepT) \times T_{step} + T_{\min} \\ Q_{C,i} &= round(rand \times stepQ_C) \times Q_{C,step} + Q_{C,\min} \end{aligned} \tag{19}$$

where $round(\cdot)$ is the rounding function; $stepT$, $stepQ_C$ are the value of tap positions and capacitor banks, respectively; T_{step}, $Q_{C,step}$ are the step size of adjustable transformers and capacitor banks, respectively.

6.3. GSAPSO Algorithm for RPO

Step 1: Initialize parameters (including population size, learning factor, inertia weight range and initial value, the maximum number of iterations, the particle velocity range, control variables and its adjustment range).

Step 2: Divide the population into four parts, and initialize the position and velocity for all the particles.

Step 3: Initialize the power flow calculation parameters and calculate the fitness value of each particle based on the fitness function (Equation (12)), and then update the optimal value of the individual and population.

Step 4: If the better particles appear (for the central particles in the second part or the chaotic particles in the fourth part), replace the poor particles in the first part with them, and update the particles.

Step 5: Update the value of inertia weight using Equations (6) to (9).

Step 6: Update the position and velocity of four part particles for the next generation, according to Equations (1) to (5).

Step 7: Calculate the fitness value of particles, and update the best position of the individual and population.

Step 8: If one of the stopping criteria (maximum iterations or fitness accuracy) is satisfied, the search procedure is stopped, else go to Step 4.

7. Simulation

The simulation experiment using standard IEEE 30 bus power system was performed to validate the GSAPSO. Detailed parameters of the system can be found in the literature [31]. The topological structure of it is shown in Figure 3, which consists of six generator buses and four transformers. To the nature of buses, bus 1 was balanced bus, bus 2, 5, 8, 11 and 13 were PV buses, and the other buses were PQ buses. Twelve control variables were involved in optimal reactive power compensation, including six generator voltages, four tap changing transformers, and two shunt compensation capacitor banks. The four branches (6–9, 6–10, 4–12 and 27–28) had transformers with tap changing. In addition, the installation location of shunt compensation capacitor banks were generally ensured at buses 10 and

24 [31]. The reference capacity of the whole power system was 100 MVA and the control variable were set as shown in Table 4.

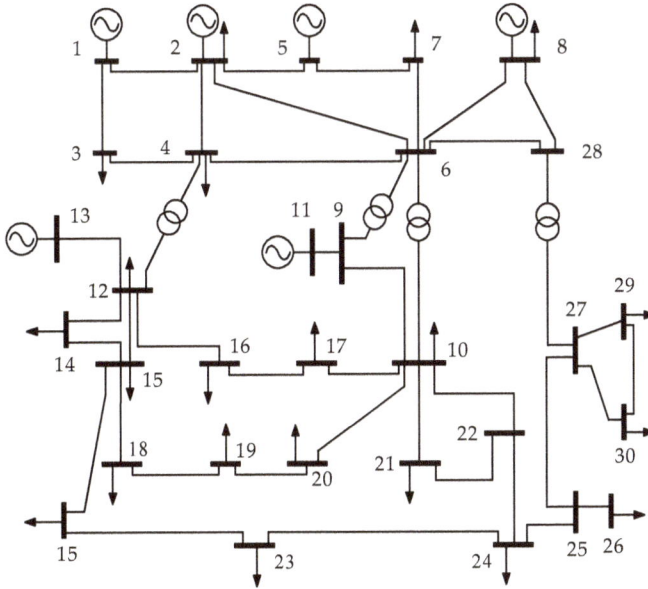

Figure 3. IEEE 30 bus power system.

Table 4. Control variable setting.

Variables	Minimum (p.u.)	Maximum (p.u.)	Step
U_G	0.9	1.1	-
U_{PQ}	0.95	1.05	-
T	0.9	1.1	0.025
C_{10}	0	0.05	0.2
C_{24}	0	0.02	0.2

The number of iterations during the calculation period was 100; the other parameter settings were the same with those stated in Section 4.2. Under the initial conditions, the generator voltage and the transformer ratio were both set to 1.0 p.u. When the compensation capacitor capacity was 0 Mvar, the initial network loss was 8.421 MW through the power flow calculation.

For evaluating the performance of GSAPSO for reactive power compensation effectively, the results were compared with PSO, LDWPSO and WPSO. 10 independent trails were carried out and the convergence accuracy took 10^{-4}. The optimal values, the worst values and the average values of various algorithms were compared. The results are presented in Table 5.

Table 5. The results with four algorithms.

Optimization Results	PSO	LDWPSO	WPSO	GSAPSO
The optimal value/MW	6.9516	6.824	6.8242	6.8239
The worst value/MW	7.2262	6.8667	6.8614	6.8699
The average value/MW	7.046	6.8342	6.8354	6.8336
The variance	7.034×10^{-3}	1.31×10^{-4}	1.21×10^{-4}	1.17×10^{-4}
Iterations of the optimal solution	99	92	77	45
Average iterations	67.3	86.8	67.2	64.6

From Table 5, the minimum network loss after the reactive power optimization calculated by the GSAPSO algorithm was 6.8239 MW, the average network loss was 6.8336 MW, the number of optimal solution iterations was 45 and the average number of iterations was 64.6. It demonstrated that the performance of GSAPSO was better than the PSO, LDWPSO and WPSO algorithms. By comparing the variances, we could see that the GSAPSO had a better convergence stability and robustness.

The GSAPSO considered the effective information provided by multiple particles in the optimization process. However, the other three algorithms only considered the role of the optimal individual in the later iteration, so that the algorithms converged slowly near the optimal value. GSAPSO increased the diversity of the population through double central particles, chaotic particles and synergistic particles and jumped out the local optimum value, which enhanced the global optimization ability.

Table 6 presents the details of control variables after optimization. GSAPSO obtaind the minimum power losses and 18.966% loss reduction, which was better than the other algorithms.

Table 6. The optimized control variables with four optimization algorithms.

Parameters	PSO	LDWPSO	WPSO	GSAPSO
U1	1.0678	1.0686	1.0686	1.0687
U2	1.0453	1.0506	1.0506	1.0506
U5	1.0227	1.0267	1.0267	1.0267
U8	1.0409	1.0381	1.0381	1.0381
U11	1.0026	1.0337	1.0099	1.0363
U13	1.0656	1.0749	1.075	1.0749
C1	40	30	40	30
C2	10	10	10	10
T1	1.025	1.1	1.025	1.05
T2	1.05	0.9	1	0.95
T3	0.975	1	1	1
T4	0.975	0.975	0.975	0.975
Ploss/MW	6.9516	6.824	6.8242	6.8239
Psave/%	17.449	18.964	18.962	18.966

Figure 4 shows the convergence curve of the average fitness value using different algorithms. From Figure 4, it was revealed that GSAPSO performed excellently compared to the other algorithms for the RPO problem. The GSAPSO took around 45 iterations to converge, whereas WPSO, LDWPSO took around 55 iterations and 75 iterations, respectively. The PSO could not converge even after 100 iterations. The voltage magnitudes of buses before and after reactive power optimization are presented in Figure 5.

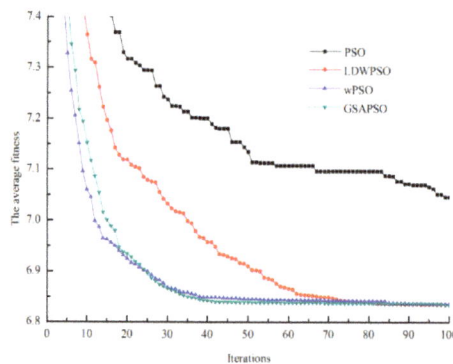

Figure 4. Convergence curves of average fitness using different algorithms.

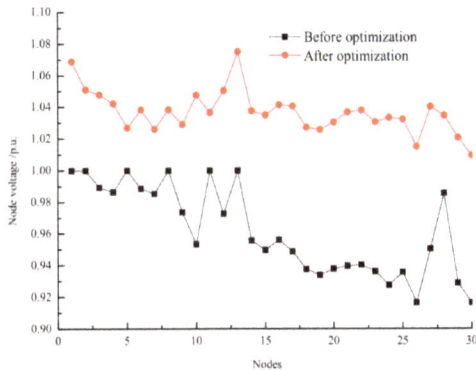

Figure 5. The bus voltage before and after optimization.

We can see that the bus voltage magnitudes after reactive power optimization were significantly higher than before optimization. Moreover, they are more stable, the minimum value is 1.0091 p.u. and all values are within the limited range.

8. Conclusions

This paper has presented the GSAPSO, which was proposed by improving the particle update modes and the inertia weight of the algorithm in the iterative process, through which the accuracy and ability of global convergence were effectively improved. The simulation experiments were conducted on four benchmark functions, which validly tested the improved PSO. The salient feature of the improved was that GSAPSO had an obviously higher convergence rate, convergence accuracy and success rate than LDWPSO and WPSO. Then the GSAPSO algorithm was applied for RPO problem on the IEEE30 bus system. The simulation results revealed that the optimal loss reduction rate could reach 18.966%, the optimal solution iteration number was 45 times, which had great improvement compared with 64.6 times, the average iteration number in others algorithms. Moreover, the minimum of bus voltage was 1.0091 p.u. and whole bus voltages were more stable. The proposed GSAPSO algorithm obtained lesser power losses and a more stable voltage compared to the others. The simulation results demonstrated that the GSAPSO had an excellent performance for solving the RPO problems.

Author Contributions: All the authors contributed to this paper on the ideas, mathematical models, the algorithms, the analysis of the simulation results and finally the editing and revision.

Funding: This research was funded by the Doctoral Scientific Research Foundation of Liaoning Province under Grant No. 201601106 and Natural Science Foundation of Liaoning under Grant No. 20170450810.

Conflicts of Interest: The authors declare no conflict of interest.

References

1. Antunes, C.H.; Pires, D.F.; Barrico, C.; Gomes, Á.; Martins, A.G. A Multi-Objective Evolutionary Algorithm for Reactive Power Compensation in Distribution Networks. *Appl. Energy* **2011**, *86*, 977–984. [CrossRef]
2. Ara, A.L.; Kazemi, A.; Gahramani, S.; Behshad, M. Optimal Reactive Power Flow Using Multi-Objective Mathematical Programming. *Sci. Iran.* **2012**, *19*, 1829–1836.
3. Zhang, Y.J.; Ren, Z. Real-Time Optimal Reactive Power Dispatch Using Multi-Agent Technique. *Electr. Power Syst. Res.* **2004**, *69*, 259–265. [CrossRef]
4. Jeyadevi, S.; Baskar, S.; Babulal, C.K.; Iruthayarajan, M.W. Solving Multiobjective Optimal Reactive Power Dispatch Using Modified Nsga-Ii. *Int. J. Electr. Power Energy Syst.* **2011**, *33*, 219–228. [CrossRef]
5. Moghadam, A.; Seifi, A.R. Fuzzy-Tlbo Optimal Reactive Power Control Variables Planning for Energy Loss Minimization. *Energy Convers. Manag.* **2014**, *77*, 208–215. [CrossRef]

6. Borghetti, A. Using Mixed Integer Programming for the Volt/Var Optimization in Distribution Feeders. *Electr. Power Syst. Res.* **2013**, *98*, 39–50. [CrossRef]

7. Mandal, B.; Roy, P.K. Optimal Reactive Power Dispatch Using Quasi-Oppositional Teaching Learning Based Optimization. *Int. J. Electr. Power Energy Syst.* **2013**, *53*, 123–134. [CrossRef]

8. Subbaraj, P.; Rajnarayanan, P.N. Optimal Reactive Power Dispatch Using Self-Adaptive Real Coded Genetic Algorithm. *Electr. Power Syst. Res.* **2009**, *79*, 374–381. [CrossRef]

9. Opoku, G. Optimal Power System Var Planning. *IEEE Trans. Power Syst.* **1990**, *5*, 53–60. [CrossRef]

10. Sun, D.I.; Ashley, B.; Brewer, B.; Hughes, A.; Tinney, W.F. Optimal Power Flow by Newton Approach. *IEEE Trans. Power Appar. Syst.* **1984**, *103*, 2864–2880. [CrossRef]

11. Lo, K.L.; Zhu, S.P. A Decoupled Quadratic Programming Approach for Optimal Power Dispatch. *Electr. Power Syst. Res.* **1991**, *22*, 47–60. [CrossRef]

12. Granville, S. Optimal Reactive Dispatch through Interior Point Methods. *IEEE Trans. Power Syst.* **1994**, *9*, 136–146. [CrossRef]

13. Iba, K. Reactive Power Optimization by Genetic Algorithm. *IEEE Trans. Power Syst.* **2002**, *9*, 685–692. [CrossRef]

14. Mao, Y.; Li, M. Optimal Reactive Power Planning Based on Simulated Annealing Particle Swarm Algorithm Considering Static Voltage Stability. In Proceedings of the 2008 International Conference on Intelligent Computation Technology and Automation (ICICTA), Hunan, China, 20–22 October 2008.

15. Varadarajan, M.; Swarup, K.S. Differential Evolutionary Algorithm for Optimal Reactive Power Dispatch. *Int. J. Electr. Power Energy Syst.* **2008**, *30*, 435–441. [CrossRef]

16. Varadarajan, M.; Swarup, K.S. Differential Evolution Approach for Optimal Reactive Power Dispatch. *Appl. Soft Comput.* **2008**, *8*, 1549–1561. [CrossRef]

17. Malekpour, A.R.; Tabatabaei, S.; Niknam, T. Probabilistic Approach to Multi-Objective Volt/Var Control of Distribution System Considering Hybrid Fuel Cell and Wind Energy Sources Using Improved Shuffled Frog Leaping Algorithm. *Renew. Energy* **2012**, *39*, 228–240. [CrossRef]

18. Sulaiman, M.H.; Mustaffa, Z.; Mohamed, M.R.; Aliman, O. Using the Gray Wolf Optimizer for Solving Optimal Reactive Power Dispatch Problem. *Appl. Soft Comput.* **2015**, *32*, 286–292. [CrossRef]

19. Singh, R.P.; Mukherjee, V.; Ghoshal, S.P. Optimal Reactive Power Dispatch by Particle Swarm Optimization with an Aging Leader and Challengers. *Appl. Soft Comput.* **2015**, *29*, 298–309. [CrossRef]

20. Niknam, T.; Firouzi, B.B.; Ostadi, A. A New Fuzzy Adaptive Particle Swarm Optimization for Daily Volt/Var Control in Distribution Networks Considering Distributed Generators. *Appl. Energy* **2010**, *87*, 1919–1928. [CrossRef]

21. El-Dib, A.A.; Youssef, H.K.M.; El-Metwally, M.M.; Osman, Z. Optimum Var Sizing and Allocation Using Particle Swarm Optimization. *Electr. Power Syst. Res.* **2007**, *77*, 965–972. [CrossRef]

22. Esmin, A.A.A.; Lambert-Torres, G.; Souza, A.C.Z.D. A Hybrid Particle Swarm Optimization Applied to Loss Power Minimization. *IEEE Trans. Power Syst.* **2005**, *20*, 859–866. [CrossRef]

23. Kennedy, J.; Eberhart, R. Particle swarm optimization. In Proceedings of the IEEE International Conference on Neural Networks, Piscataway, NJ, USA, 27 November–1 December 1995; pp. 1942–1948.

24. Kannan, S.M.; Renuga, P.; Kalyani, S.; Muthukumaran, E. Optimal Capacitor Placement and Sizing Using Fuzzy-De and Fuzzy-Mapso Methods. *Appl. Soft Comput.* **2011**, *11*, 4997–5005. [CrossRef]

25. Zad, B.B.; Hasanvand, H.; Lobry, J.; Vallée, F. Optimal Reactive Power Control of DGs for Voltage Regulation of MV Distribution Systems Using Sensitivity Analysis Method and PSO Algorithm. *Int. J. Electr. Power Energy Syst.* **2015**, *68*, 52–60.

26. Achayuthakan, C.; Ongsakul, W. TVAC-PSO based optimal reactive power dispatch for reactive power cost allocation under deregulated environment. In Proceedings of the IEEE Conference on Power Energy and Industry Applications, Calgary, AB, Canada, 26–30 July 2009; pp. 1–9.

27. Vaisakh, K.; Rao, P.K. Optimal reactive power allocation using PSO-DV hybrid algorithm. In Proceedings of the India Conference, Kanpur, India, 11–13 December 2008.

28. Tang, K.Z.; Liu, B.X. Double Center Particle Swarm Optimization Algorithm. *J. Comput. Res. Dev.* **2012**, *49*, 1086–1094.

29. Arasomwan, M.A.; Adewumi, A.O. On the Performance of Linear Decreasing Inertia Weight Particle Swarm Optimization for Global Optimization. *Sci. World J.* **2013**, *2013*, 1–12. [CrossRef] [PubMed]

30. Ren, Z.H.; Wang, J. New Adaptive Particle Swarm Optimization Algorithm with Dynamically Changing Inertia Weight. *Comput. Sci.* **2009**, *36*, 227–229.

31. Lee, K.Y.; Park, Y.M.; Ortiz, J.L. A United Approach to Optimal Real and Reactive Power Dispatch. *IEEE Power Eng. Rev.* **2010**, *PER–5*, 42–43.

energies

Article

Implementation and Comparison of Particle Swarm Optimization and Genetic Algorithm Techniques in Combined Economic Emission Dispatch of an Independent Power Plant

Shahbaz Hussain [1], Mohammed Al-Hitmi [1], Salman Khaliq [2,*], Asif Hussain [3] and Muhammad Asghar Saqib [4]

[1]   Department of Electrical Engineering, Qatar University, P.O. Box 2713 Doha, Qatar;
       shahbazpec@yahoo.com (S.H.); m.a.alhitmi@qu.edu.qa (M.A.-H.)
[2]   Intelligent Mechatronics Research Center, Korea Electronics Technology Institute (KETI),
       Gyeonggi-do 13509, Korea
[3]   Department of Electrical Engineering, University of Management and Technology, Lahore 54792, Pakistan;
       asif.hussain@umt.edu.pk
[4]   Department of Electrical Engineering, University of Engineering and Technology, Lahore 54890, Pakistan;
       saqib@uet.edu.pk
*   Correspondence: salman.khaliq@hotmail.com; Tel.: +82-10-2695-6234

Received: 11 March 2019; Accepted: 7 May 2019; Published: 28 May 2019

Abstract: This paper presents the optimization of fuel cost, emission of NO_X, CO_X, and SO_X gases caused by the generators in a thermal power plant using penalty factor approach. Practical constraints such as generator limits and power balance were considered. Two contemporary metaheuristic techniques, particle swarm optimization (PSO) and genetic algorithm (GA), have were simultaneously implemented for combined economic emission dispatch (CEED) of an independent power plant (IPP) situated in Pakistan for different load demands. The results are of great significance as the real data of an IPP is used and imply that the performance of PSO is better than that of GA in case of CEED for finding the optimal solution concerning fuel cost, emission, convergence characteristics, and computational time. The novelty of this work is the parallel implementation of PSO and GA techniques in MATLAB environment employed for the same systems. They were then compared in terms of convergence characteristics using 3D plots corresponding to fuel cost and gas emissions. These results are further validated by comparing the performance of both algorithms for CEED on IEEE 30 bus test bed.

Keywords: economic load dispatch; emission dispatch; combined economic emission/environmental dispatch; particle swarm optimization; genetic algorithm; penalty factor approach

---

1. Introduction

The primary objective of an independent electric power producer is to generate electricity at the minimum possible cost. The most expensive commodity in a thermal power plant is the fuel used for the generators to produce electricity. Hence, the focus is on minimizing the production cost; which can be achieved by dispatching the committed generators in the most economical way possible without violating the generators and system electrical constraints and ratings. Moreover, environmental regulatory authorities also impose certain limits on all gas emission sources because of the alarming situation of pollution in the past few decades of many regions around the world [1]. Therefore, a simultaneous minimization of both fuel cost and emission is the obvious way to address those challenges; hence, the idea of a combined economic emission dispatch (CEED) emerged.

Combined dispatch is an efficient and economical solution for decreasing both fuel cost and emission in a thermal power plant without the need to modify the existing system. The simulation gives flexibility to the operator to set the output of generators to achieve a fuel cost benefit for the company and emission allowed by the environmental regulatory authorities. There are many optimization techniques being employed to solve multiobjective problems like CEED. Conventional methods are based on mathematical iterative search which are accurate but time consuming. The nonconventional methods are naturally inspired and give better, if not the best, solution in lesser time as compared to conventional methods. From these artificial-intelligence-based methods, a process of hybridization is going on to accumulate the qualities of individual methods into their hybrids counterparts. This categorization is shown in Figure 1 [2].

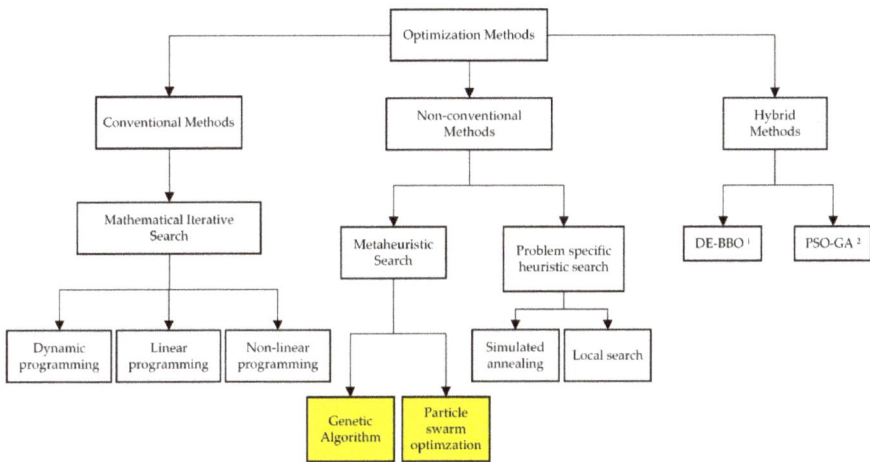

Figure 1. Optimization methods. [1] Differential Evolution-Biogeography Based Optimization. [2] Particle Swarm Optimization-Genetic Algorithm.

Many modern artificial intelligence (AI) techniques based on the theory of Darwinian evolution of biological organisms, e.g., genetic algorithm (GA), and social behaviors of species, e.g., particle swarm optimization (PSO), have been invented. They have been very successful concerning results and regarding dealing with the complexities occurred in formulating such problems. Particle swarm optimization and genetic algorithm are the two leading methods from AI area and are being exploited widely in every discipline including power systems [3–6]. Their hybrid versions are also reported in the literature [7–9], where qualities of both are combined to solve a particular problem. They have also been employed individually on other problems [10,11] where PSO was found to have better performance than that of GA.

In recently reported literature, E. Gonçalves et al. solved nonsmooth CEED using a deterministic approach with improved performance and Pareto curves [1]. They included valve point loading effect but the same approach has to be extended considering other constraints like network losses and prohibited operating zones. H. Liang et al. tackled the multiobjective combined dispatch by developing a hybrid and improved version of bat algorithm [12] and implemented on large scale systems considering power flow constraints. The conducted dispatch was static based on conventional energy resources. B. Lokeshgupta et al. proposed a combined model of multiobjective dynamic economic and emission dispatch (MODEED) and demand side management (DSM) technique using multiobjective particle swarm optimization (MOPSO) algorithm and validated their results via three different cases studied on a six unit test system [13] and can be extended towards distributed generation in microgrids.

In reported literature [14–16], in most cases, only one technique has been used to solve CEED on IEEE test cases, and their results have been compared with previous work. In Table 1, the survey of [17] is summarized to conclude that GA is better for low-power systems while PSO outperforms in the case of high-power systems. However, what happens when these two leading metaheuristic techniques are employed together to the same system? This work deals with this novel idea, hence, both PSO and GA will be implemented individually on CEED of an independent power plant (IPP) situated in Pakistan considering all gases (NO_X, CO_X, and SO_X) for various load demands. The results will be of great importance as the data of an actual IPP will be utilized and they will also grade the performance of PSO and GA for CEED of an IPP. In order to validate the results, a conventional IEEE 30 bus system is also considered. This work will contribute the results of combined dispatch of fuel and gas emissions (economic and environmental aspects, respectively) carried out on a power plant with two leading algorithms (PSO and GA) using MATLAB and then comparing their performance in terms of better solution, convergence characteristics (3D plots), and computation time. The implementation, comparison, and convergence characteristics of PSO and GA employed for the same systems are the main contribution of this work to the field. These characteristics were compared with each other with 3D plots to analyze the better solution yielded by the two algorithms applied for CEED of the same systems in MATLAB environment.

Table 1. Best optimization methods employed in different systems with regard to cost, emission, and time

No.	System	Cost-effective	Emission-effective	Time-effective
1.	1 power unit, 2 cogeneration units, and 1 heat unit	Harmony search and genetic algorithm (HSGA)	X	Cuckoo search algorithm (CSA)
		Nondominated sorting genetic algorithm (NSGA-II)	Nondominated sorting genetic algorithm (NSGA-II)	Nondominated sorting genetic algorithm (NSGA-II)
2.	1 power unit, 3 cogeneration units, and 1 heat unit	HSGA	X	Gravitational search algorithm (GSA)
3.	4 thermal generators, 2 cogeneration units, and 1 heat unit	GSA	X	Effective cuckoo search algorithm (ECSA)
		Grey wolf optimization (GWO)	GWO	GWO
4.	13 power units, 6 cogeneration units, and 7 heat units	Exchange market algorithm (EMA)	X	Modified particle swarm optimization (MPSO)
5.	26 power units, 12 cogeneration units, and 10 heat units	MPSO	X	MPSO

2. Problem Formulation

CEED comprises two objective functions (cost and emission) that are to be minimized. The fuel cost and emission of a generator can be represented as a quadratic function of the generator's real power [18]. Hence, for N running generators in a plant, the total fuel cost (FC) and emission of a single gas (E_g), respectively, are given in Equations (1) and (2).

$$FC = \sum_{i=1}^{N} F_i(P_i) = \sum_{i=1}^{N} \left(a_i P_i^2 + b_i P_i + c_i\right) \tag{1}$$

$$E_g = \sum_{i=1}^{N} E_i(P_i) = \sum_{i=1}^{N} \left(\alpha_i P_i^2 + \beta_i P_i + \gamma_i\right) \tag{2}$$

where a_i, b_i, and c_i are cost coefficients; α_i, β_i, and γ_i are emission coefficients of unit i out of N generators. They are computed by the following:

- Getting the heat rate curves and emission reports of operational generators from the plant.
- Then, calculating and arranging their fuel costs and emissions corresponding to their active powers in tabular form.
- Applying the quadratic curve fitting technique on these data points to get cost and emission coefficients.

This multiobjective optimization problem is converted to a single objective function of total cost (TC) by imposing a penalty (h_g) on the emission of G gases to convert them into emission cost (EC).

$$TC = FC + EC = FC + \sum_{g=1}^{G} h_g \times E_g \tag{3}$$

There are different types of penalty factors; their benefits and drawbacks are thoroughly discussed in [19]. The min–max penalty factor of Equation (4) is used in this work because of its superiority reported in [19] over the others.

$$h_i = \frac{F_i(P_{i,min})}{E_i(P_{i,max})} = \frac{a_i P_{i,min}^2 + b_i P_{i,min} + c_i}{\alpha_i P_{i,max}^2 + \beta_i P_{i,max} + \gamma_i} \tag{4}$$

where h_i for each generator for every gas is calculated and all are sorted in ascending order, then starting from the smallest h_i, $P_{i,max}$ of corresponding generator is added until $\sum P_{i,max} \geq P_D$, the h_i at this stage is selected as penalty factor h_g of that gas for the given load demand.

For achieving this optimization, the N generators will be dispatched with various combinations of output powers but each combination must conform to two mandatory constraints. First of all, no unit should violate its limits for producing output power (P_i), and secondly the total generation (P_G) should meet the load demand (P_D) and transmission line losses (P_L) [18].

$$P_{i,min} \leq P_i \leq P_{i,max} \tag{5}$$

$$P_G = \sum_{i=1}^{N} P_i = P_D + P_L \tag{6}$$

The generator limits constraint is satisfied by initializing each unit's power within prescribed limits and then constantly checking the violation. If a unit crosses its limit, then its output is set to that limit. The second power balance constraint is accounted for by letting algorithm to find optimal powers for $N-1$ generators and setting the power (P_N) of last generator N also called slack generator to:

$$P_N = (P_D + P_L) - \sum_{i=1}^{N-1} P_i \tag{7}$$

If losses are ignored, then the power of slack generator will reduce to:

$$P_N = P_D - \sum_{i=1}^{N-1} P_i \tag{8}$$

3. Implementation of PSO to CEED

J. Kennedy and R. Eberhart proposed this method in 1995 after observing and modeling the social interaction within bird flocks and fish schools for searching food [20]. The particles in such swarms move to attain optimal objective (food) based on their personal (*pbest*) and swarm's (*gbest*) best experiences. Each particle is a valid solution to the problem, and hence, its dimension is that of problem space. The position of each particle keeps on updating by its current velocity, particle's best position, and swarm's best position until the optimal solution is discovered. The velocity and position of the particle is given by Equations (9) and (10).

$$v_j^{k+1} = \mu^k v_j^k + c_1 r_1 \left(pbest_j - x_j^k \right) + c_2 r_2 \left(gbest - x_j^k \right) \tag{9}$$

$$x_j^{k+1} = x_j^k + v_j^{k+1} \tag{10}$$

where j is particle counter, k is iteration counter, c_1 and c_2 are acceleration coefficients, r_1 and r_2 are random numbers in the range of 0–1, $pbest_j$ is the best position of particle based on its personal knowledge, $gbest$ is the best position of particle based on group knowledge, and μ is the inertia weight given by Equation (11).

$$\mu^k = \mu_{max} - \left(\frac{\mu_{max} - \mu_{min}}{iter_{max}} \right) k \tag{11}$$

The performance of classical PSO has been made a lot better by working on its various parameters and by using different search strategies for updating the particle's position [18]. Hence, different variants of PSO has been introduced, such as WIPSO and TVAC-PSO, in which inertia weight μ is improved and acceleration coefficients c_1 and c_2 are timely varied [21], MRPSO [22] in which particle's position is changed by using moderate random and chaotic search techniques, respectively. It has been observed that the most straightforward and efficient way of making classical PSO more effective is to use the constriction factor approach (CFA) [23], in which particle's velocity (9) is multiplied by a parameter called constriction factor (*CF*) given by Equation (12).

$$CF = \frac{2}{\left| 2 - \varphi - \sqrt{\varphi^2 - 4\varphi} \right|} \tag{12}$$

where $\varphi = c_1 + c_2$ and $\varphi > 4$. The PSO algorithm for CEED, with corresponding flowchart in Figure 2, is implemented in the following steps:

1. Input values of fuel coefficients, generators limits, emission coefficients, load demand, maximum iterations, number of particles, acceleration coefficients, and inertia weight's maxima-minima.
2. Randomly initialize power outputs (position) of $N - 1$ generators within their limits and change in these powers (velocity) for all particles.
3. Calculate the power of slack generator from Equation (8) to meet the power balance condition of Equation (6). P_N should also be within its unit's limits. If any unit out of N surpasses its boundary at any stage throughout the algorithm, it is set to the limit which it has broken.
4. Initialize *pbest* and *gbest* to infinity and find penalty factors of all considered gases.
5. Find fuel cost *FC*, emission of gases E_g, emission cost *EC* with total cost *TC* for all particles from Equations (1)–(3), respectively.
6. Update *pbest* of each particle with its total cost if former is greater than latter. The minimum *pbest* out of all particles is stored as *gbest* if its present value is smaller than its previous value. The generators' powers corresponding to *pbest* and *gbest* are stored in separate matrices and are also modified on every update.
7. Calculate inertia weight from Equation (11), find new velocities and positions of $N-1$ generators for all particles form Equations (9) and (10), and keep them within units' limits in case of violation.
8. For slack generator N, calculate its power from Equation (8) and this should also be in limits. If violation is done, set this power to the limit crossed and start changing the output from first generator until Equations (5) and (6) are satisfied.
9. If algorithm is converged, continue to next step, go to step 5 otherwise.

10. Print neatly the results of optimal solution (*gbest*) including powers of all generators, line losses, fuel cost, emission of gases, penalty factors, emission cost, and total cost for the given load demand.

4. Implementation of GA to CEED

D.E. Goldberg gave the basic theory for design and analysis of genetic algorithms based on the concept of biological evolution in 1988–1989 [24], and later, J.H. Holland established it systematically as a fact. In genetic algorithms, the problem variables (output powers) are coded into binary strings. Each string is a valid solution to the problem and hence its length should be comparable to the problem space (number of generators N). The bits reserved logically for single generator's power (*genbits*) in a string is given in Equation (13).

$$2^{genbits} \geq Max[P_{1,max}, P_{N,max}] \tag{13}$$

The length of the binary string (*strlen*) will be $N \times genbits$. Each individual (string) has a fitness value in the range 0–1 which basically relates that individual to the one having maximum fitness in that population. The fitness function should be linked to the objective under discussion contrary to the constraints of the objective function which should be dealt by external checks. Considering this fact and the remarkable communication within the group by PSO parameters *pbest* and *gbest*, a new fitness function different from the one reported in [3] and [5] for *j*th individual out of P individuals is given by (14).

$$fit_j = 1 - \left(\frac{pbest_j - gbest}{Max[pbest_1, pbest_P] - gbest} \right) \tag{14}$$

The initial population is randomly generated, keeping in view the generators' limits but the next generations are produced by selection, crossover, and mutation performed on the present one (powers corresponding to *pbest*). Selection is basically making a mating pool of fitter strings from present population based on the natural principle of "survival of the fittest." It is usually done by the concept of roulette-wheel. No new string is formed in selection phase. The greater the fitness of a string, the greater portion of the wheel's circumference it will occupy and the greater chance it will get to copy into mating pool. The wheel is spun P times to select a population of good parents for producing off springs by crossover and mutation.

Crossover is performed on two parents of selected population to produce two off springs. There are three types of crossover: one point, multi-point, and uniform, explained in Table 2.

Table 2. The types of crossover.

Item	One Point	Multipoint	Uniform
Parent 1	000 00000	00 00 00 00	00 00 00 00
Parent 2	111 11111	11 11 11 11	11 11 11 11
Site/Mask	3	2, 4, 6	01 01 10 10
Child 1	000 11111	00 11 00 11	01 01 10 10
Child 2	111 00000	11 00 11 00	10 10 01 01

Crossover site is selected randomly and the probability of crossover (p_c) is usually taken higher. In this work, one point crossover is performed on selected strings of mating pool. Finally, mutation is performed on the children produced after crossover which is just flipping of the child's bit at mutation site selected randomly. Its probability (p_m) is usually taken lower, e.g., if child is (11 11 11 11) and mutation site is 4, then the mutated child will be (11 10 11 11). This journey of producing next generations continues until the optimum solution is found.

The GA algorithm for CEED, with corresponding flowchart in Figure 3, is implemented in the following steps:

1. Input values of fuel coefficients, generators limits, emission coefficients, load demand, maximum iterations, number of individuals, *genbits* from Equation (13), *strlen*, p_c, and p_m.

2. Randomly initialize power outputs of $N-1$ generators within their limits for all individuals.

3. Calculate the power of slack generator from Equation (8) to meet power balance condition of Equation (6). P_N should also be within its unit's limits. If any unit out of N surpasses its boundary at any stage throughout the algorithm, it is set to the limit which it has broken.

4. Initialize *pbest* and *gbest* to infinity and find penalty factors of all considered gases.

5. Find fuel cost *FC*, emission of gases E_g, emission cost *EC* with total cost *TC* for all particles from Equations (1)–(3), respectively.

6. Update *pbest* of each individual with its total cost if former is greater than latter. The minimum *pbest* out of all particles is stored as *gbest* if its present value is smaller than its previous value. The generators' powers corresponding to *pbest* and *gbest* are stored in separate matrices and are also modified on every update.

7. Calculate fitness function for all individuals from Equation (14). Code the output powers to binary strings, perform the three genetic operators (selection, crossover, and mutation), and again decode them to output powers.

8. Keep all outputs within units' limits in case of violation. For slack generator N, calculate its power from Equation (8), and this should also be in limits. If violation is done, set this power to the limit crossed and start changing the output from first generator until Equations (5) and (6) are satisfied.

9. If algorithm is converged, continue to next step, go to step 5 otherwise.

10. Print neatly the results of optimal solution (*gbest*) including powers of all generators, line losses, fuel cost, emission of gases, penalty factors, emission cost, and total cost for the given load demand.

5. Simulation Results

Combined economic emission dispatch using PSO and GA for 500 iterations were implemented on MATLAB on six generators of IEEE 30 bus system and eight committed units (gas turbines) of an IPP in Pakistan for load demands of 1500 and 2000 MW, and 500 and 700 MW, respectively. The initial parameters set in both algorithms were:

PSO: Particles = 10, $\mu_{max} = 0.9$, $\mu_{min} = 0.4$, $c_1 = 2.05$, $c_2 = 2.05$, $\varphi = 4.1$, and CF = 0.7298
GA: Individuals = 10, $p_c = 0.96$, and $p_m = 0.033$

The solutions with average operating time (t) were selected out of 50 trials for comparison between PSO and GA.

5.1. IEEE 30 Bus System

The data for fuel cost and emission coefficients were taken from [5]. Transmission line losses were not accounted for while all three gases (NO_X, CO_X, and SO_X) were considered; their penalty factors were calculated using Equation (4) and the procedure following this equation, for load demands of 1500 and 2000 MW given as:

$P_D = 1500$ MW: $h_{NOX} = 3.1669$, $h_{COX} = 0.1221$, and $h_{SOX} = 0.9182$
$P_D = 2000$ MW: $h_{NOX} = 5.7107$, $h_{COX} = 0.1307$, and $h_{SOX} = 0.9850$

The results are summarized in Table 3 (all powers are in MW, emissions in kg/h, costs in $/h, and time in seconds) while the convergence characteristics of both algorithms, with respect to both objectives (fuel cost and emission) for $P_D = 1500$ MW and $P_D = 2000$ MW, are shown in Figures 4 and 5, respectively.

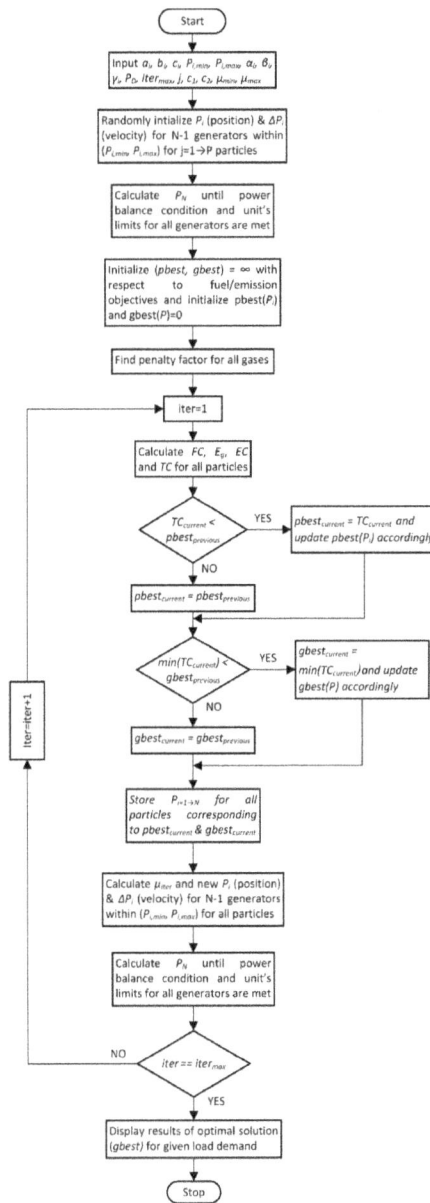

Figure 2. Implementation of combined economic emission dispatch (CEED) using particle swarm optimization (PSO) algorithm.

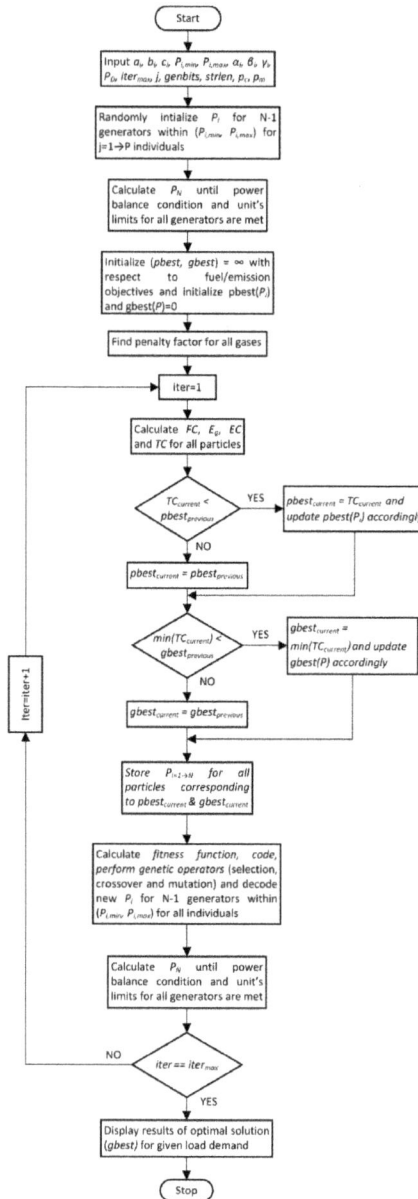

Figure 3. Implementation of CEED using genetic algorithm (GA) algorithm.

Table 3. The real-time simulation results of PSO and GA for IEEE 30 bus system with P_D = 1500 MW and P_D = 2000 MW.

Case A		P1	P2	P3	P4	P5	P6	E_{NOX}	E_{COX}	E_{SOX}	E	EC	FC	TC	t
P_D = 1500 MW	PSO	195.79	256.55	381.25	81.69	381.85	202.27	1719.67	39,626.51	9623.04	50,969.21	19,121.26	14,827.57	33,948.83	0.1326
	GA	196	255	351	79	416	203	1731.52	40,049.21	9579.83	51,360.56	19,170.74	14,835.04	34,005.78	0.4528
P_D = 2000 MW	PSO	256.32	320.57	541.95	133.1	500	248.06	2540.68	73,738.76	13,601.05	89,880.5	37,543.01	19,445.29	56,988.3	0.1227
	GA	268	287	562	127	500	256	2537.8	75,505.33	13,450.06	91,493.19	37,608.68	19,485.44	57,094.12	0.4706

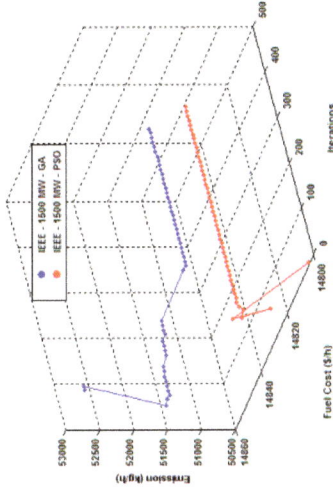

Figure 4. The convergence characteristics of PSO and GA with regard to fuel cost and emission for IEEE 30 bus system with P_D = 1500 MW.

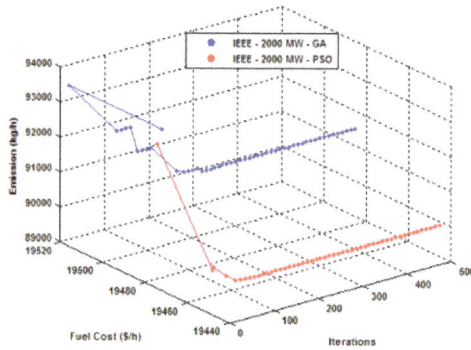

Figure 5. The convergence characteristics of PSO and GA with regard to fuel cost and emission for IEEE 30 bus system with $P_D = 2000$ MW.

5.2. Pakistani IPP

This thermal power plant comprises of 15 generating units out of which 10 are multi-fuel-fired gas turbines and remaining five are steam turbines with an overall capacity of 1600 MW. The combined dispatch was performed on eight gas turbines as the remaining two are uneconomical and mostly turned off. Steam turbines take exhaust of gas turbines to operate, hence they were not considered as they do not take any direct fuel.

The data calculated and used for the dispatch of all units is given in Appendix A. Transmission line losses were ignored because IPP's main concern is generation capacity which they have to produce and supply to national grid, and SO_X gas were not accounted for because of the unavailability of sufficient data. Remaining two gases (NO_X and CO_X) were considered; their penalty factors were calculated using Equation (4) and the procedure following this equation, for load demands of 500 and 700 MW given as:

$P_D = 500$ MW: $h_{NOX} = 1.5751$, $h_{COX} = 101.1369$
$P_D = 700$ MW: $h_{NOX} = 1.7218$, $h_{COX} = 123.8797$

The results are summarized in Table 4 (all powers are in MW, emissions in mg/Nm^3, costs in 10^3 \$/h, and time in seconds) while the convergence characteristics of both algorithms, with respect to both objectives (fuel cost and emission) for $P_D = 500$ MW and $P_D = 700$ MW, are shown in Figures 6 and 7, respectively.

Table 4. The real-time simulation results of PSO and GA for Pakistani independent power plant (IPP) with $P_D = 500$ MW and $P_D = 700$ MW.

Case B		P1	P2	P3	P4	P5	P6	P7	P8	E_{NOX}	E_{COX}	E	EC	FC	TC	t
$P_D = 500$ MW	PSO	32.5	32.5	100	90.87	83.68	100	25	35.44	2512.49	40.04	2552.53	76	117.1	193.1	0.1531
	GA	33	32.5	32	92	96	100	64	50.5	2624.22	43.32	2667.54	80.82	121.6	202.42	0.8031
$P_D = 700$ MW	PSO	130	130	100	90.83	83.82	100	25	40.35	3093.41	48.93	3142.35	108.09	158.5	266.59	0.1509
	GA	130	130	100	87	96	100	25	32	3110.72	55.41	3166.13	115.1	158.4	274.4	0.5411

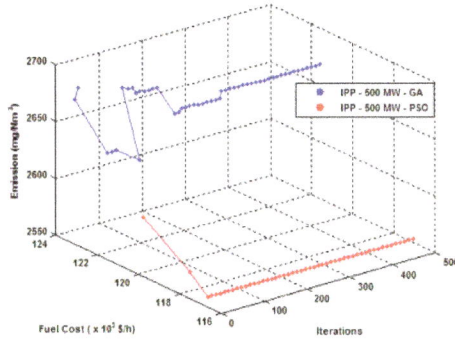

Figure 6. The convergence characteristics of PSO and GA with regard to fuel cost and emission for Pakistani IPP with $P_D = 500$ MW.

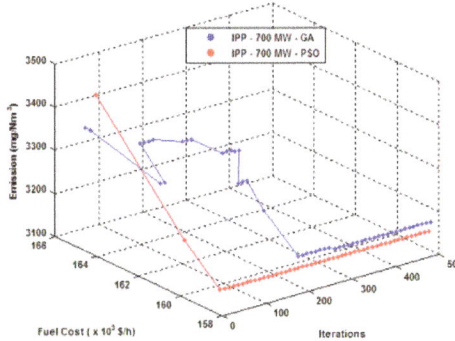

Figure 7. The convergence characteristics of PSO and GA with regard to fuel cost and emission for Pakistani IPP with $P_D = 700$ MW.

6. Results Discussion

In Table 3, the power output of all six generators was presented as the sum of which is equal to the load demand (1500 and 2000 MW) for both algorithms (PSO and GA). Then, the emissions of all three gases (NO_X, CO_X, SO_X) from these generators were calculated and added (E) to convert into cost (EC) using penalty factor approach. The fuel cost (FC) is calculated from the generators' powers and is added to the emission cost to achieve total cost (TC) of the combined dispatch. In the end, the given time (*t*) represents the computational time of the algorithm for that particular load demand and PSO produces the optimal solution 4 times quicker than that of GA for both load demands. Figure 4 shows the 3D plot of fuel cost and emission with regard to iterations for 1500 MW. It shows that PSO starts with lower fuel cost and emission, and converges quickly on the optimal solution as compared to GA which starts with higher fuel cost and emission and takes some time to converge. Figure 5 shows the similar 3D plot but for a load demand of 2000 MW in which PSO gets lost for a small time period in non-optimized region but quickly converges on the solution with lower fuel cost and emission, while GA converges in steps on to a solution which is not better than PSO.

Table 4 gives the output power of eight gas-fired generators and this total generation is equal to the load demand (500 and 700 MW) for both cases (PSO and GA). From the generators' powers, the fuel cost (FC) and emission of two gases (NO_X and CO_X) were calculated using quadratic relation of generator's output to fuel cost and emission, respectively. The emission was then totaled (E) and converted to cost (EC) by imposing penalty factor. These two costs (related to fuel and emission) were then added to get total cost (TC) of the combined dispatch of the power plant. In the end, *t* is the time taken by each algorithm to carry out the dispatch for a particular load demand. It is noteworthy that PSO is 4–7 times faster than its GA counterpart for both load demands. Figure 6 shows the 3D plot of fuel cost and emission with regard to iterations for 500 MW. In this figure, PSO starts with higher fuel cost but in few iteration, it converges on the optimal solution while GA remains stuck in non-optimized region. Figure 7 shows the similar 3D plot but for a load demand of 700 MW. In this simulation, both PSO and GA starts from nearby solutions, but then, PSO converges very quickly while GA struggles up to 300 iterations and then converges to a nearer solution of PSO, but still, this solution is not better than the one provided by PSO.

7. Conclusions

The combined dispatch of plant generators with respect to fuel and gas emissions was carried out using PSO and GA in MATLAB. The results demonstrate that PSO outperforms GA for the combined dispatch in terms of achieving lower fuel cost, lower emission, fast convergence, and lesser simulation time. This was validated for both IEEE 30 bus system and for an independent power plant with simulation, 3D plots, and thorough discussion. The work has successfully implemented PSO and GA algorithms for CEED of the same systems. The simulation results are compared and tabulated for different load demands. 3D plots were discussed to highlight the convergence characteristics of PSO and GA with respect to fuel cost and gas emissions. In future studies, the combined dispatch will be made more realistic, by including some other practical constraints like valve point loading, ramp rate limits, prohibited operating zones of generators, transmission line losses etc. The algorithms, PSO and GA, can be made more efficient by studying and improving their performance parameters.

Author Contributions: Conceptualization and Formal analysis, S.H.; Funding acquisition, M.A.-H.; Investigation, S.H.; Methodology, S.H., S.K. and A.H.; Project administration, M.A.-H. and M.A.S.; Resources, M.A.-H.; Software, S.H. and A.H.; Supervision, M.A.-H. and M.A.S.; Writing—original draft, S.H.; Writing—review & editing, S.K.

Acknowledgments: The publication charges of this article were funded by Qatar National Library (QNL), Doha, Qatar.

Conflicts of Interest: The authors declare no conflict of interest.

Appendix A

Table A1. Fuel cost coefficients.

Unit	a	b	c
1	−0.053809	29.524	−0.59888
2	−0.019141	25.719	−10.238
3	−0.14098	36.796	0.0017735
4	−0.089013	32.088	0.094472
5	−0.024246	26.81	−7.0972
6	−0.050478	28.482	−13.301
7	−0.042498	30.046	−8.1329
8	−0.098058	33.612	−3.2653

Generators Limits: $32.5 \leq P_{1-2} \leq 130$ and $25 \leq P_{3-8} \leq 100$.

Table A2. NO$_X$ emission coefficients.

Unit	α	β	γ
1	−0.033656	8.438	0.060408
2	−0.03745	8.8286	−0.038644
3	−0.02653	6.9845	−0.064953
4	−0.008363	3.9176	125.51
5	0.0064342	4.2814	0.036778
6	0.0061679	4.1367	−0.081828
7	−0.0076829	5.8882	−0.04459
8	0.0064342	4.2814	0.036778

Table A3. CO$_X$ emission coefficients.

Unit	α	β	γ
1	0.0005961	−0.036862	2
2	-7.5165×10^{-6}	0.03361	0.0041904
3	0.00032592	−0.0069586	5.02
4	0.10945	−19.783	899.51
5	0.033333	−5.5833	235.7
6	−0.0016337	0.16934	6
7	0.0016643	−0.13608	9.8
8	0.0022332	−0.19187	12.4

References

1. Gonçalves, E.; Roberto Balbo, A.; da Silva, D.N.; Nepomuceno, L.; Baptista, E.C.; Soler, E.M. Deterministic approach for solving multi-objective non-smooth Environmental and Economic dispatch problem. *Int. J. Electr. Power Energy Syst.* **2019**, *104*, 880–897.
2. Mahdi, F.P.; Vasant, P.; Kallimani, V.; Watada, J.; Siew Fai, P.Y.; Abdullah-Al-Wadud, M. A holistic review on optimization strategies for combined economic emission dispatch problem. *Renew. Sustain. Energy Rev.* **2018**, *81*, 3006–3020. [CrossRef]
3. Mishra, S.K.; Mishra, S.K. A comparative study of solution of economic load dispatch problem in power systems in the environmental perspective. *Proc. Comput. Sci.* **2015**, *48*, 96–100. [CrossRef]
4. Kothari, D.P. Power system optimization. In Proceedings of the 2012 2nd National Conference on Computational Intelligence and Signal Processing (CISP), Guwahati, Assam, India, 2–3 March 2012.
5. AlRashidi, M.R.; El-Hawary, M.E. Emission-economic dispatch using a novel constraint handling particle swarm optimization strategy. In Proceedings of the 2006 Canadian Conference on Electrical and Computer Engineering, Ottawa, ON, Canada, 7–10 May 2006.
6. Bora, T.C.; Mariani, V.C.; dos Santos Coelho, L. Multi-objective optimization of the environmental-economic dispatch with reinforcement learning based on non-dominated sorting genetic algorithm. *Appl. Therm. Eng.* **2019**, *146*, 688–700. [CrossRef]
7. Ghamisi, P.; Benediktsson, J.A. Feature selection based on hybridization of genetic algorithm and particle swarm optimization. *IEEE Geosci. Remote Sens. Lett.* **2015**, *12*, 309–313. [CrossRef]
8. Jiang, H.; Zhang, Y.; Xu, H. Optimal allocation of cooperative jamming resource based on hybrid quantum-behaved particle swarm optimisation and genetic algorithm. *IET Radar Sonar Navig.* **2016**, *11*, 185–192. [CrossRef]
9. Wang, J.; Zhang, F.; Liu, F.; Ma, J. Hybrid forecasting model-based data mining and genetic algorithm-adaptive particle swarm optimisation: A case study of wind speed time series. *IET Renew. Power Gener.* **2016**, *10*, 287–298. [CrossRef]
10. Ohatkar, S.N.; Bormane, D.S. Hybrid channel allocation in cellular network based on genetic algorithm and particle swarm optimisation methods. *IET Commun.* **2016**, *10*, 1571–1578. [CrossRef]
11. Villarroel, R.D.; García, D.F.; Dávila, M.A.; Caicedo, E.F. Particle swarm optimization vs genetic algorithm, application and comparison to determine the moisture diffusion coefficients of pressboard transformer insulation. *IEEE Trans. Dielectr. Electr. Insul.* **2015**, *22*, 3574–3581. [CrossRef]

12. Liang, H.; Liu, Y.; Li, F.; Shen, Y. A multiobjective hybrid bat algorithm for combined economic/emission dispatch. *Int. J. Electr. Power Energy Syst.* **2018**, *101*, 103–115. [CrossRef]
13. Lokeshgupta, B.; Sivasubramani, S. Multi-objective dynamic economic and emission dispatch with demand side management. *Int. J. Electr. Power Energy Syst.* **2018**, *97*, 334–343. [CrossRef]
14. Güvenç, U.; Sönmez, Y.; Duman, S.; Yörükeren, N. Combined economic and emission dispatch solution using gravitational search algorithm. *Sci. Iranica* **2012**, *19*, 1754–1762.
15. Jebaraj, L.; Venkatesan, C.; Soubache, I.; Christober Asir Rajan, C. Application of differential evolution algorithm in static and dynamic economic or emission dispatch problem: A review. *Renew. Sustain. Energy Rev.* **2017**, *77*, 1206–1220. [CrossRef]
16. Niu, Q.; Zhang, H.; Wang, X.; Li, K.; Irwin, G.W. A hybrid harmony search with arithmetic crossover operation for economic dispatch. *Int. J. Electr. Power Energy Syst.* **2014**, *62*, 237–257. [CrossRef]
17. Nazari-Heris, M.; Mohammadi-Ivatloo, B.; Gharehpetian, G. A comprehensive review of heuristic optimization algorithms for optimal combined heat and power dispatch from economic and environmental perspectives. *Renew. Sustain. Energy Rev.* **2018**, *81*, 2128–2143. [CrossRef]
18. Mahor, A.; Prasad, V.; Rangnekar, S. Economic dispatch using particle swarm optimization: A review. *Renew. Sustain. Energy Rev.* **2009**, *13*, 2134–2141. [CrossRef]
19. Krishnamurthy, S.; Tzoneva, R. Investigation on the impact of the penalty factors over solution of the dispatch optimization problem. In Proceedings of the 2013 IEEE International Conference on Industrial Technology (ICIT), Cape Town, South Africa, 25–28 February 2013.
20. Kennedy, J. Particle swarm optimization. *Encycl. Mach. Learn.* **2010**, 760–766. [CrossRef]
21. Vu, P.; Dinh Le, L.; Vo, N.; Tlusty, J. A novel weight-improved particle swarm optimization algorithm for optimal power flow and economic load dispatch problems. In Proceedings of the IEEE PES T&D 2010, New Orleans, LA, USA, 19–22 April 2010.
22. Singh, N.; Kumar, Y. Economic load dispatch with environmental emission using MRPSO. In Proceedings of the 2013 3rd IEEE International Advance Computing Conference (IACC), Ghaziabad, India, 22–23 February 2013.
23. Dasgupta, K.; Banerjee, S. An analysis of economic load dispatch using different algorithms. In Proceedings of the 2014 1st International Conference on Non Conventional Energy (ICONCE 2014), Kalyani, India, 16–17 January 2014.
24. Goldberg, D.E.; Holland, J.H. Genetic algorithms and machine learning. *Mach. Learn.* **1988**, *3*, 95–99. [CrossRef]

energies

MDPI

Article

Recognition and Classification of Incipient Cable Failures Based on Variational Mode Decomposition and a Convolutional Neural Network

Jiaying Deng, Wenhai Zhang and Xiaomei Yang *

College of Electrical Engineering and Information Technology, Sichuan University, Chengdu 610065, China; 2017223035198@stu.scu.edu.cn (J.D.); zhangwh2079@scu.edu.cn (W.Z.)
* Correspondence: yangxiaomei@scu.edu.cn; Tel.: +86-028-8546-6818

Received: 16 April 2019; Accepted: 20 May 2019; Published: 25 May 2019

Abstract: To avoid power supply hazards caused by cable failures, this paper presents an approach of incipient cable failure recognition and classification based on variational mode decomposition (VMD) and a convolutional neural network (CNN). By using VMD, the original current signal is decomposed into seven modes with different center frequencies. Then, 42 features are extracted for the seven modes and used to construct a feature vector as input of the CNN to classify incipient cable failure through deep learning. Compared with using the original signals directly as the CNN input, the proposed approach is more efficient and robust. Experiments on different classifiers, namely, the decision tree (DT), K-nearest neighbor (KNN), BP neural network (BP) and support vector machine (SVM), and show that the CNN outperforms the other classifiers in terms of accuracy.

Keywords: incipient cable failure; VMD; feature extraction; CNN

1. Introduction

The development process of cable failure is usually divided into three stages: the partial discharge period, incipient failure period and permanent failure period [1]. Due to defect formation, corrosion or aging of the insulating layer, a series of partial discharge pulses first appear in the cable, forming electrical or water branches. With further deterioration, these branches would evolve into incipient failures accompanied by arcing. Incipient failures advance progressively after the first occurrence to the point of irreversible permanent failure [2]. The occurrence of incipient cable failure (also known as self-cleaning failure) is uncertain, and the current, which is very small when failure occurs, is typically not sufficient to trigger the safety protection of traditional overcurrent detection devices [3,4]. At the same time, because of the similarity with other disturbances, e.g., overcurrents caused by transformer inrush current, constant impedance and capacitor switching failures, it is relatively difficult to recognize incipient cable failures accurately. Therefore, it is of great significance for the stable operation of power grid to achieve an approach that can effectively recognize and classify incipient cable failures and to complete cable maintenance in time before incipient failures become permanent failures.

Due to the uncertainty of the transmission environment and the effects of noise, the actual cable signals usually contain a large amount of interference information. Especially under high noise conditions, the features of incipient cable failure are so weak that it is difficult to directly recognize them. Therefore, it is necessary to perform signal preprocessing before classification. At present, many researchers have proposed different approaches for the recognition and classification of cable failures. The recognition methods are mainly divided into fault detection methods based on hardware design [5–8] and feature extraction based on time-frequency analysis [9–14]. In [5], a hardware relay implementation in power systems is presented, which by using a proposed syntactic pattern [6], reads each peak of the examined signals to detect existing faults. The effectiveness of the proposed relay is determined by the user setting suitable thresholds. In [7], the waveform is monitored for alarming distortion by using the proposed amplitude tracking algorithm in a field programmable gate array (FPGA). In [8], Romano et al. presented an antenna sensor to measure partial discharge. The results show that the instrument can diagnose the health of the insulation system quickly and accurately. Obviously, fault detection based on hardware design is fast and real-time, but it also depends on hardware algorithms, and the threshold setting and portability of the devices must be considered.

Feature extraction methods based on time-frequency analysis include short-time Fourier transform (STFT) [9], wavelet transform (WT) [10], and empirical mode decomposition (EMD) [11]. In [12], an arc fault recognition systems that combine STFT with hardware devices is described; the system checks whether the harmonic components in the frequency domain exceed a threshold value that is set according to limited testing loads in the lab. The STFT overcomes the shortcoming of frequency domain time information loss of the Fourier transform, and realizes local time-frequency analysis of the signal. However, the size of the window function cannot change with frequency, which is not conducive to analyzing time-varying signals. In [13], Sidhu et al. decomposed the original current signals via WT, and the overcurrent disturbance was initially recognized by using the energy and root mean square value at the fault location; then, incipient cable failure was identified by adjusting the thresholds. Chao et al. [14] detected incipient cable failure based on the overcurrents and the WT modulus maximum of the sum of single-end sheath currents. When the magnitude of the sum of sheath currents exceeds the theoretical value, an overcurrent condition can be diagnosed. Then, one can set the threshold based on the normal signal for the WT modulus maxima of the sum of sheath currents to determine whether the overcurrents are generated by incipient failure. WT inherits and develops the idea of STFT localization, and it overcomes the shortcomings of window fixation. However, its decomposition results depend too much on the choice of wavelet basis. In [15], the instantaneous frequency component of the partial discharge signal of the cable joint was obtained via EMD, and then the 3D (time-frequency-energy) Hilbert spectrum was calculated. Finally, partial discharge types were evaluated by using a neural network. EMD not only breaks through the limitations of FT, but also does not have the problem of preselecting wavelet basis function such as the wavelet transform. It has good time-frequency resolution and adaptability. However, this algorithm lacks a mathematical foundation. In particular, modal aliasing problems are easily generated during the decomposition process.

In this paper, a feature extraction method based on time-frequency analysis of variational mode decomposition (VMD) and mathematical statistics is used. VMD was proposed by Dragomiretskiy et al. in 2014 [16]; it is a complete nonrecursive adaptive and quasi-orthogonal signal decomposition method that decomposes multicomponent nonstationary signal into several components with limited bandwidth. The essence of VMD is that multiple sets of Wiener filters are updated directly in the Fourier domain; thus, VMD has a more solid theoretical basis and robustness than EMD [17,18]. At the same time, 42 features are extracted by the method of mathematical statistics, which solves the feature selection problem and improves the calculation efficiency.

Fault classification methods can be categorized into machine learning algorithms and deep learning algorithms. Machine learning algorithms, including decision tree (DT), support vector machine (SVM), K-nearest neighbor (KNN), and artificial neural network (ANN), have been widely used in power systems [19–23]. Recently, researchers have conducted studies of traditional algorithms for different problems and further improved the classification accuracy [24–26]. However, these classifiers still have some defects, such as a weak correlation between data, a tendency to overfit and a limited capacity for the real-time processing of large data. A deep learning model is a network structure that updates the parameters of multiple hidden layers via a gradient descent algorithm, which overcomes the shortcomings of machine learning algorithms and solves the problems of data utilization, gradient diffusion and trapping in local optima [27,28]. The representative structures are the deep belief network (DBN) [29], the recurrent neural network (RNN) [30,31] and the convolutional neural network (CNN) [32]. The DBN algorithm can reflect the similarity of the same type of data itself, but since its generation model does not care about the optimal classification surface, the classification accuracy is usually not as high as that of the discrimination model. The RNN has time memory ability through recursive method, but it requires more training parameters and is prone to gradient dissipation or gradient explosion.

This paper uses the CNN algorithm; as one of the deep learning models, its multilevel structure of local perception and value sharing can accommodate high-dimensional data more accurately and quickly. Due to the characteristics of self-learning, especially the results are stable in the field of image and signal recognition, and there is no additional feature engineering requirement [33].

This paper presents a new approach for incipient cable failure recognition and classification based on VMD feature extraction and CNN classification (Figure 1). Via VMD, features of each decomposed mode for the original signal are extracted, and feature vector is input into the CNN to recognize incipient cable failure. The paper is organized as follows: in Section 2, the VMD principle, mathematical model and CNN algorithm are described. In Section 3, experimental data are generated by constructing a model of incipient cable failure and overcurrent interference in PSCARD/EMTDC. Next, the complete experimental results are shown. Compared with traditional machine learning algorithms (DT, KNN, BP, and SVM), the proposed approach not only can recognize incipient cable failure state well but also has good noise immunity. In Section 4, the produced results are commented and the limitations of the research are expounded. In the last section, the conclusions are summarized.

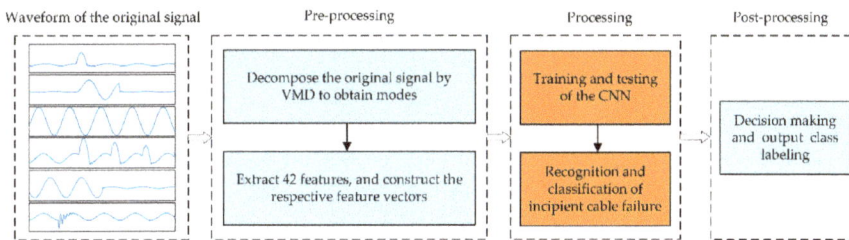

Figure 1. Main structure of the proposed incipient cable failure recognition and classification process.

2. Methods

2.1. VMD

VMD is used to decompose the nonstationary cable signal $x(t)$ into a number of subcomponents (known as modes), with each mode having a finite bandwidth corresponding to the center frequency. Thus, VMD aims to minimize the sum of the bandwidth estimate of each mode. To adaptively extract modes and their central frequencies, an optimization problem is constructed as follows:

First, let $x(t)$ be predecomposed into K modes, denoted as u_k $(k = 1, \cdots, K)$, i.e., $\sum_{k=1}^{K} u_k = x(t)$. To acquire the unilateral frequency spectrum, the associated analysis signal of u_k is formed by using the Hilbert transform as follows:

$$\left[\delta(t) + \frac{j}{\pi t} \right] * u_k(t), \tag{1}$$

where $\delta(t)$ is the Dirac delta function, j is an imaginary number, and * is the convolutional operation.

Then, for each mode u_k, its frequency spectrum is modulated to the baseband through multiplying the corresponding analysis signal by the exponential $e^{-j\omega_k t}$, where ω_k is the center frequency of u_k.

Next, to adaptively extract u_k and ω_k, by using the L^2-norm of the demodulated signal's gradient, the constrained variational optimization problem is constructed as follows:

$$\begin{cases} \min\limits_{\{u_k\},\{\omega_k\}} \left\{ \sum\limits_{k=1}^{K} \left\| \partial_t \left[\left(\delta(t) + \frac{j}{\pi t} \right) * u_k(t) \right] e^{-j\omega_k t} \right\|_2^2 \right\} \\ s.t. \sum\limits_{k=1}^{K} u_k = x(t) \quad (k = 1, \cdots, K) \end{cases} \tag{2}$$

where $\partial_t(\cdot)$ is the partial derivative and $\|\cdot\|_2$ is the L^2-norm.

Finally, to solve for u_k and ω_k in (1), by using the augmented Lagrange multiplier method [34], the constrained variational problem (2) is transformed into the following unconstrained variational problem:

$$L(\{u_k\}, \{\omega_k\}, \lambda) = \alpha \sum_{k=1}^{K} \left\| \partial_t \left[\left(\delta(t) + \frac{j}{\pi t} \right) * u_k(t) \right] e^{-j\omega_k t} \right\|_2^2 + \left\| x(t) - \sum_{k=1}^{K} u_k(t) \right\|_2^2 + \left(\lambda(t), x(t) - \sum_{k=1}^{K} u_k(t) \right) \tag{3}$$

where λ is Lagrange multiplier and α is the parameter to balance variational term (the first term) and data-fidelity constraint term (the second term), so as to impart good reconstruction accuracy to the signal.

Then, u_k, ω_k and λ in (2) are iteratively and alternately updated by using the alternating direction multiplier method (ADMM) in the Fourier domain [15]. In the nth iteration, u_k^{n+1} and ω_k^{n+1} are calculated in the Fourier domain as follows:

$$\hat{u}_k^{n+1}(\omega) = \frac{\hat{x}(\omega) - \sum\limits_{i \neq k}^{K} \hat{u}_i(\omega) + \frac{\hat{\lambda}(\omega)}{2}}{1 + 2\alpha(\omega - \omega_k)^2}, \tag{4}$$

and:

$$\omega_k^{n+1} = \frac{\int_0^\infty \omega |\hat{u}_k(\omega)|^2 d\omega}{\int_0^\infty |\hat{u}_k(\omega)|^2 d\omega}, \tag{5}$$

respectively. The Lagrange multiplier λ is updated as follows:

$$\hat{\lambda}^{n+1}(\omega) = \hat{\lambda}^n(\omega) + \tau \left[\hat{x}(\omega) - \sum_k \hat{u}_k^{n+1}(\omega) \right]. \tag{6}$$

where τ is the time step of the dual ascent. Once $\sum_k \|\hat{u}_k^{n+1} - \hat{u}_k^n\|^2 / \|\hat{u}_k^n\|^2 < \varepsilon (\varepsilon > 0)$ is satisfied, the iteration process is over, and ω_k is obtained. Furthermore, the mode u_k in the time domain is obtained as the real part of the inverse Fourier transform of $\hat{u}_k$.

2.2. Feature Extraction

VMD decomposes the input signal into multifrequency signals, which is helpful for effectively distinguishing incipient cable failure from that of other disturbances. Obviously, the decomposed multilayer signals have more data. If input them into the CNN for classification directly would increase the difficulty of selecting the network parameters and the training time be too long. Thus, to achieve the aim of efficiently training the CNN, features extracted from decomposed VMD modes are used as inputs of the CNN below. For each VMD mode with N sample points, denoted as $u_k[n]$ ($1 \leq n \leq N$), six features are extracted as follows:

F1: *F1* represents the peak-to-peak value, i.e., the range of signal amplitude variation. An observation of six types of signals indicates that *F1* of multicycle cable failure is several times that of other signals. Therefore, this feature plays a key role in distinguishing between multicycle cable failure from the range of amplitudes of different signals. For the k-th mode, *F1* is given by:

$$F1_k = max(u_k[n]) - min(u_k[n]). \tag{7}$$

F2: This feature is the root mean square (RMS), which represents the effective value of the periodic signal. Because the RMS is the squared average calculated over one cycle, the feature helps to separate all multicycle failure signals from other types. Hence, *F2* is an important reference in the recognition of multicycle cable failure, normal signals, transformer inrush disturbance and constant impedance failure. For the *k*th mode, *F2* is given by:

$$F2_k = \sqrt{\frac{1}{N} \sum_n u_k^2[n]}. \tag{8}$$

F3: The feature *F3* represents the number of zero crossings, which is used to identify the oscillatory characteristics and noise of the stationary signal for each mode. For the *k*th mode, *F3* is computed as follows:

$$F3_k = \sum_n |sgn(u_k[n]) - sgn(u_k[n-1])|/2. \tag{9}$$

Here, sgn(·) is a symbolic function. Taking into account the duration of different nonstationary signals at the zero position, *F3* reflects the state of the waveform, where the signal crosses zero for each mode, which is helpful to distinguish the overcurrent disturbance, normal signal and noise.

F4: The mode relative energy ratio represents the contribution of the energy of each mode to the total energy of the original signal. For the *k*th mode, this ratio is denoted as:

$$F4_k = \frac{\sum_n u_k^2[n]}{\sum_k \sum_n u_k^2[n]}. \tag{10}$$

The main energy of different signals is distributed over different decomposition modes, e.g., the energy of the normal signals is mainly distributed over the first one decomposition mode, and the latter modes contain only DC components. According to this feature, incipient cable failure, normal signals and overcurrent disturbance can be distinguished accurately. *F5:* The instantaneous amplitude envelope of each mode is used to distinguish between short- and long-duration variations (such as subcycle failure, transformer inrush disturbance and constant impedance failure). For the *k*-th mode, *F5* is defined using a sliding window of length *s* as follows:

$$F5_k = \sqrt{\frac{1}{s} \sum_{n=m}^{m+s-1} u_k^2[n]}. \tag{11}$$

Here, $m = 1, 2 \cdots (N-s+1)$ represents the number of sliding windows.

F6: The center frequency (ω_K) is selected as the sixth feature. Since the VMD process is implemented in the frequency domain, this condition eliminates the need for additional calculation of the frequency domain information of each mode. *F6* is used to distinguish the DC component, fundamental frequency, and oscillation of different modes.

Based on the abovementioned features, the feature vector for the *k*-th mode can be obtained as follows:

$$\mathbf{FV_k} = [F1_k, F2_k, F3_k, F4_k, F5_k, F6_k]. \tag{12}$$

In summary, assuming that the original signal is decomposed into *K* modes via VMD, a $1 \times 6K$ feature vector **F** can be composed as follows:

$$\mathbf{F} = \begin{bmatrix} \mathbf{FV_1} & \mathbf{FV_2} & \cdots & \mathbf{FV_K} \end{bmatrix} = \begin{bmatrix} F1_1 & \cdots & F6_1 & F1_2 & \cdots & F6_2 & \cdots\cdots & F6_K & \cdots & F6_K \end{bmatrix}. \tag{13}$$

Then, **F** is input to the CNN as all feature information of the original signal.

2.3. CNN

The CNN is constructed by analogy to animal visual perception. Its structure of multilayer perceptron can effectively reduce preprocessing, and it can also analyze high-dimensional data. Hence, the CNN has exhibited excellent performance in the field of image and video recognition and signal processing.

The CNN consists of an input layer, an output layer and a hidden layer. As the most important part of the whole network, the hidden layers of the CNN include convolutional layers, an activation function, downsampling layers and a fully connected layer, as shown in Figure 2. In this paper, the classification of incipient cable failure from the normal signal and the overcurrent disturbance signal is achieved through a typical CNN structure with two convolutional layers.

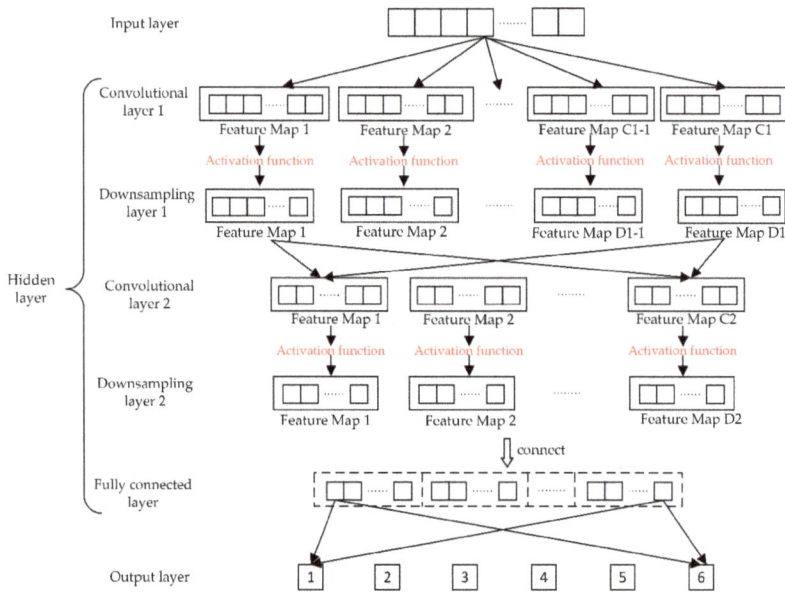

Figure 2. Structure of the CNN with two convolutional layers.

The convolutional layer is used to further extract high-level implicit features from the low-level input features using the convolutional operation. Let the feature vector **F** be the input of the CNN; the system obtains the feature map via the convolution operation and then outputs this map to the downsampling layer through the activation function. The convolution process of the feature vector **F** and the convolutional kernel **p** with the size of $1 \times G$ is as follows:

$$\mathbf{C}(1,n) = \mathbf{F} * \mathbf{p} + b = \sum_{g=1}^{G} [\mathbf{F}(1, n+g-1) \cdot \mathbf{p}(1,g)] + b. \tag{14}$$

where b is the bias, $1 \leq g \leq G$ and $n = 1, 2, \cdots, (6K - g + 1)$. **C** is the output of the convolutional layers.

Activation functions are usually used after convolution to help express more complex feature mapping. The entire process of convolution and activation can be given as follows:

$$\mathbf{C}_j^l = f\left(\sum \left(\mathbf{F}^l * \mathbf{p}_j^{l+1}\right) + b^l\right). \tag{15}$$

Here, f represents the sigmoid activation function, i.e.:

$$f(x) = \frac{1}{1 - e^{-x}}, \tag{16}$$

C_j^l is the result of the j-th feature mapping of the lth layer. $\mathbf{F}^l$ is the input vector of the l-th layer, b^l is the bias of the lth layer, and $\mathbf{p}_j^{l+1}$ is the weight coefficient of the j-th convolutional kernel of the $(l+1)$-th layer.

The convolutional results after activation are used as the input of the next downsampling layer of the neural network. The downsampling layer, also known as the pooling layer, serves to progressively reduce the size of the feature map to reduce the amount of computation in the network, thus avoiding overfitting. Suppose that the size of the feature map $\mathbf{C}$ is $1 \times M$ and the size of the downsampling area is $1 \times d$. The mean downsampling process is as follows:

$$\mathbf{D}(1, z) = \frac{1}{d} \sum_{i=1}^{d} \mathbf{C}(1, d(z-1) + i), \tag{17}$$

where $1 \leq z \leq \frac{M}{d}$. $\mathbf{D}$ is the result of the downsampling.

In general, the convolutional layer and the downsampling layer are alternately constructed. In this paper, two convolutional layers and downsampling layers are used.

The fully connected layer, as the last part of the hidden layer, connects the output of the downsampling layer to a one-dimensional row vector. Finally, the normalized exponential function applies the softmax function as the output layer to obtain the class label of classification.

3. Results

Figure 3 shows the flowchart of recognition and classification of incipient cable failure based on VMD and CNN.

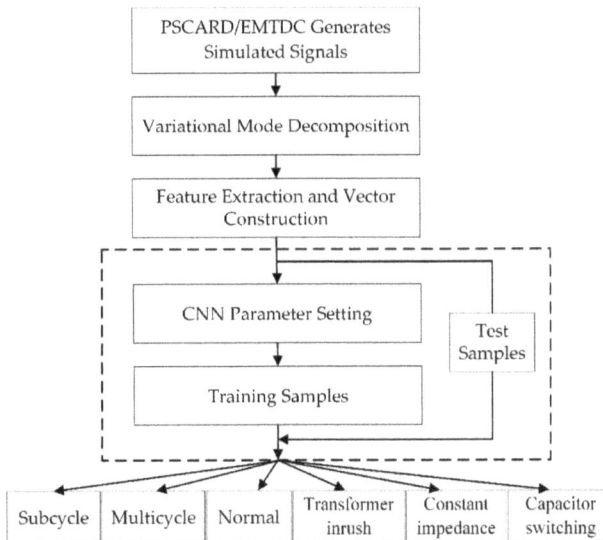

Figure 3. The flowchart based on VMD and CNN.

3.1. Data Acquisition

In this paper, all data are obtained in PSCARD/EMTDC. (Subcycle and multicycle failure have the same symbol, but the duration time is different.) The 25 kV circuit model of the cable current signal is displayed in Figure 4, and the sampling frequency is set to 10 kHz.

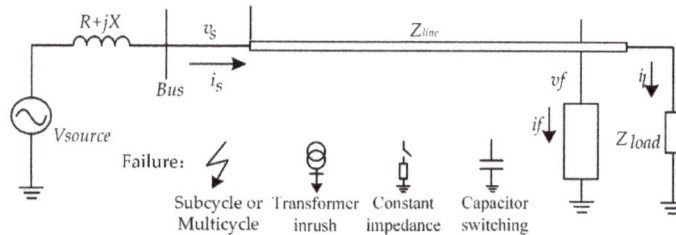

Figure 4. Simulation model of incipient cable failure and the overcurrent interference signals.

Incipient cable failure generally consists of single-phase ground failure, which can easily lead to phase-to-phase failure. Typical failure types mainly include subcycle incipient failure and multicycle incipient failure. As in references [4–7], the characteristics of incipient cable failure can be expressed as follows: (a) the failure is of short duration or low current amplitude; (b) the failure occurs at the peak of voltage; (c) a subcycle incipient failure lasts for 1/4 cycle, and when the current passes zero, the failure disappears automatically; multicycle incipient failure general lasts for 1–4 cycles, and when the arc disappears, the failure disappears automatically. In this paper, six simulation signals are used, namely, subcycle incipient cable failure, multicycle incipient cable failure, normal signal, transformer inrush disturbance, constant impedance failure and capacitor switching disturbance, as shown in Figure 5. All simulation samples are generated based on the actual interference waveform. The failure location is random, and signals of the same class exhibit differences.

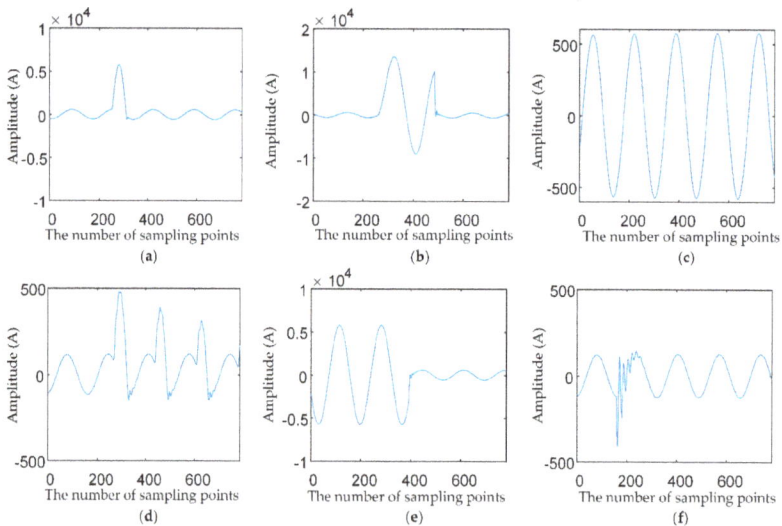

Figure 5. Current waveforms of incipient cable failure, normal signal and overcurrent interference: (a) Subcycle incipient cable failure (b) Multicycle incipient cable failure (c) Normal signal (d) Transformer inrush disturbance (e) Constant impedance failure (f) Capacitor switching disturbance.

3.2. Experimental Analysis

In the VMD process, it is necessary to set parameters. For the characteristics of the original signals, the frequency distribution is wide and includes more low-frequency information, so the parameters of VMD use the following settings:

(i) Moderate bandwidth constraint α: This parameter affects the bandwidth of the decomposed signal. For the signals with a wide frequency range, α is generally set in the range of a few hundred, while for signal content in a small frequency range, α is kept in the range of tens of thousands. According to the frequency range of the six types of signals in this experiment, $\alpha = 2000$ is used.

(ii) Noise-tolerance τ: This parameter affects the equal constraint of the Lagrange multiplier λ on reconstruction. In general, if accurate reconstruction is not required under high noise, τ can be set to 0 such that λ is 0.

(iii) The number of decomposed modes K: VMD needs to preset the number of decompose modes K. If K is too small, the decomposed modes are too few, and all the decomposition modes cannot be captured; while if the value of K is too large, the interfering signal will be overdecomposed such that the center frequencies of modes will be mixed. By repeated observation in the experiment, the maximum center frequency and the minimum center frequency of the six types of signals are nearly constant when $K = 7$, and there is no large fluctuation with increasing decomposition mode, so $K = 7$ is used.

Due to space limitations, only Figures 6 and 7 show the results and spectra of VMD for multicycle incipient cable failure and transformer inrush disturbance; the analysis process of other signals is similar. It is obvious that the waveforms of each mode corresponding to different signals differ greatly. Regarding spectra, two types of signals composed of low-frequency and intermediate-frequency content are better decomposed by VMD, so that it is fully prepared for subsequent feature extraction.

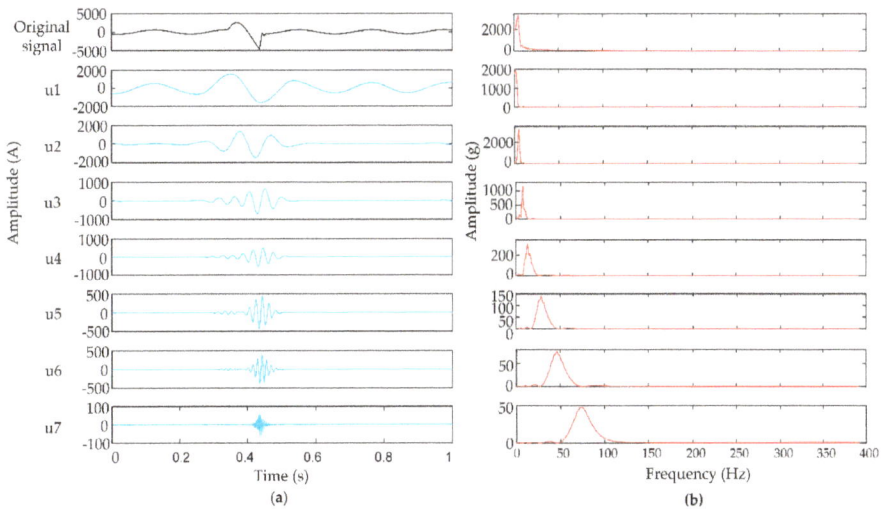

Figure 6. Results and spectra of VMD for multicycle incipient cable failure: (a) Modes; (b) Frequency spectra.

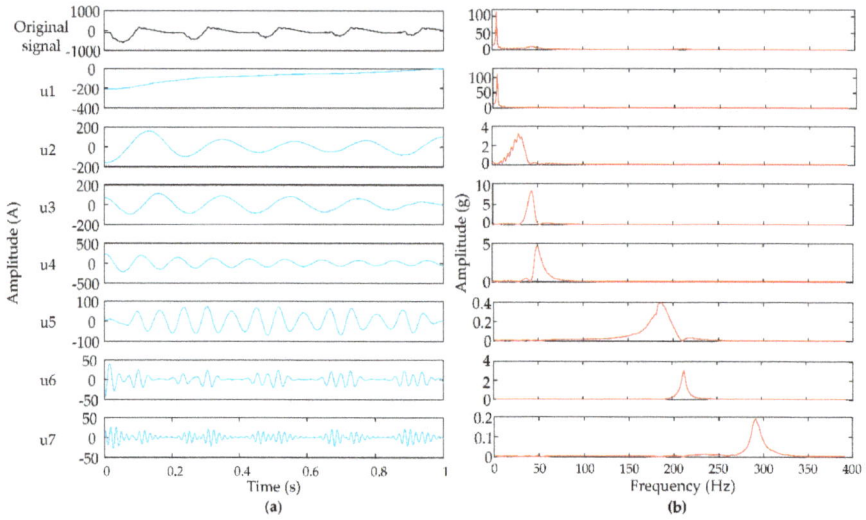

Figure 7. Results and spectra of VMD for transformer inrush disturbance: (**a**) Modes; (**b**) Frequency spectra.

Figure 8 displays feature vectors $\mathbf{FV_k}$ obtained via VMD-based stratification extraction method for a total of 42 features. As the figure shows, in each mode, the same features of different signals are mostly distinct. For example, the first feature of each vector represents $F1$, and there are significant differences in the u1–u7 modes of different signals. $F6$ is the last term of each feature vector; although it is not changed in the u1–u3 modes of different signals, there is a certain difference in u3–u7. Therefore, these feature vectors can be used as the basis for different signal diagnostics and thus input into the CNN for recognition and classification.

The structure of the CNN has an important impact on the network performance and output accuracy. Considering the computational complexity and accuracy, the parameters that are more effective for simulation data are finally selected in this paper, as indicated in Table 1. At the same time, the learning rate is set to 1, with 60 samples for each training and more than 30,000 iterations. In theory, the more iterations of the deep learning network, the higher the accuracy. In Figure 9, the impact of iteration number on classification results is shown.

Table 1. The parameters of the CNN model.

Layer Name	Kernel Size	Output Size
Input layer		$1 \times 42 \times 1$
Convolutional layer 1	1×3	$1 \times 40 \times 18$
Downsampling layer 1	1×2	$1 \times 20 \times 18$
Convolutional layer 2	1×6	$1 \times 15 \times 14$
Downsampling layer 2	1×5	$1 \times 3 \times 14$
Fully connected layer		$1 \times 42 \times 1$
Output layer		6 classes

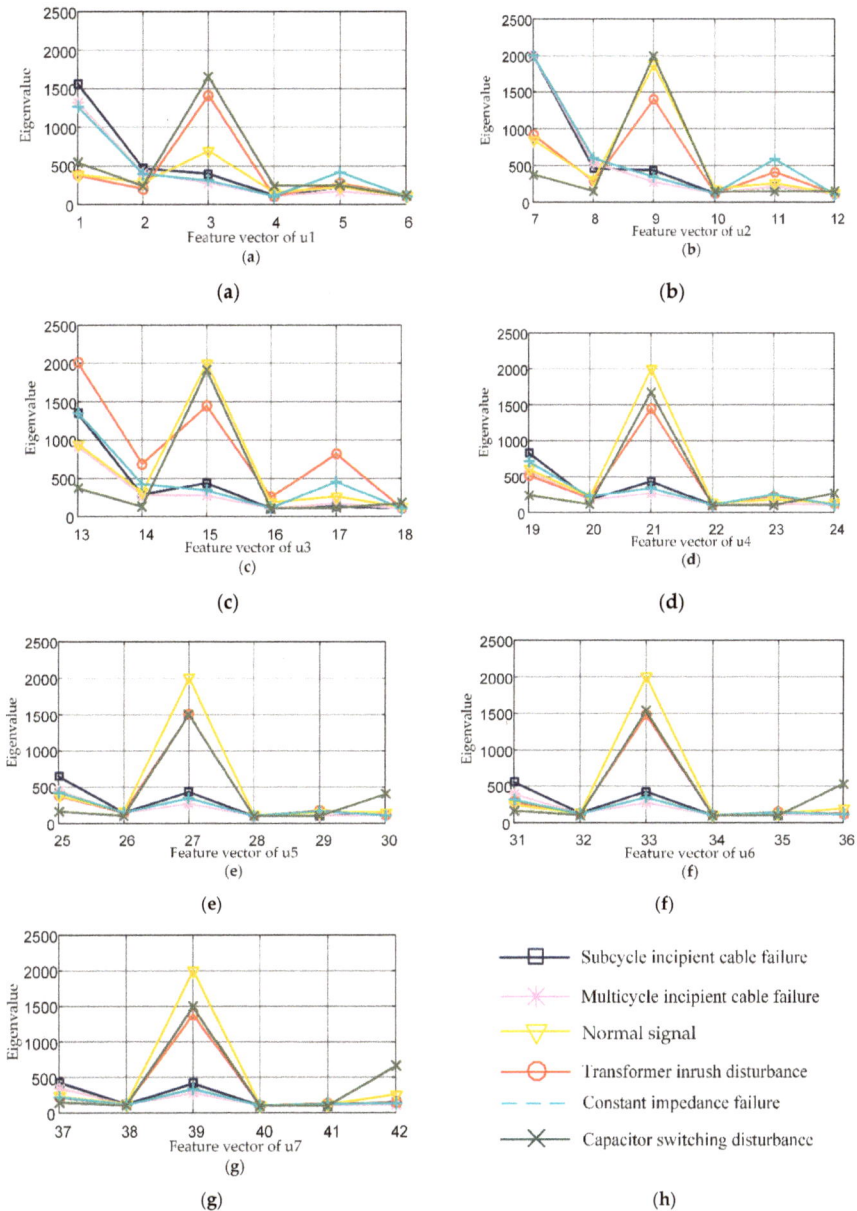

Figure 8. Feature extraction based on VMD: (**a**)–(**g**) denotes the respective feature vector **FV**$_k$ of the u1–u7 decomposed modes. (**h**) denotes the representation of different signal types.

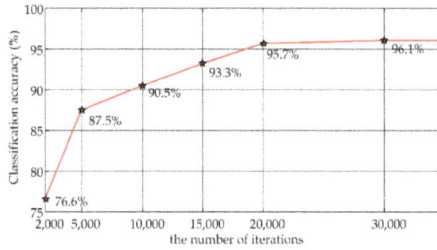

Figure 9. Impact of the iteration number on the results.

The distribution of training samples for the entire experiment is reported in Table 2.

Table 2. The number of samples.

Type	The Number of Samples	Training Samples	Test Samples
Subcycle incipient failure	2800	2100	700
Multicycle incipient failure	2800	2100	700
Normal signal	2800	2100	700
Transformer inrush	2800	2100	700
Constant impedance	2800	2100	700
Capacitor switching	2800	2100	700

For the final classification results of the subcycle incipient cable failure, multicycle incipient cable failure, normal signal, transformer inrush disturbance, constant impedance failure and capacitor switching disturbance, the performance of the model is presented in the form of a confusion matrix. As shown in Figure 10, the overall prediction accuracy is 96.1%. The rows represent the output labels, the columns represent the target labels, and the diagonal represents the number of samples correctly classified for each class. The results show that the most common mistake is to classify multicycle incipient failure as constant impedance failure; the recognition rate of multicycle incipient failure is 91.4%. The recognition rate of the normal signals is all correct, achieving 100%.

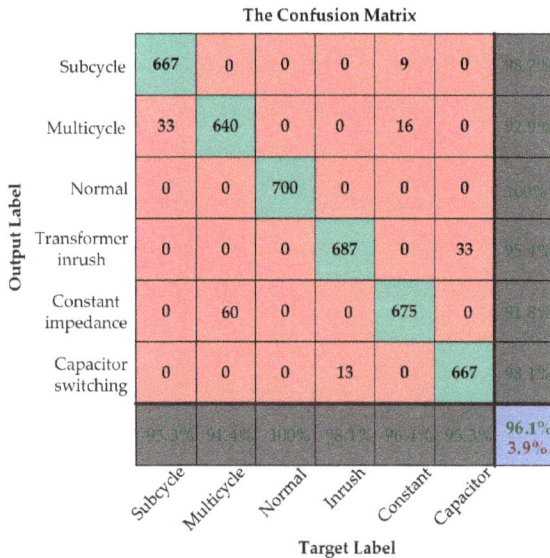

Figure 10. The results of the classification confusion matrix based on VMD and CNN.

As reported in Table 3, in order to verify the validity of the proposed approach, the original signals are used directly as the CNN input to compare with feature extraction based on VMD.

Table 3. Comparison between VMD feature extraction and the original signals.

Preprocessing	CNN Training Time (s)	Classification Accuracy (%)
Feature extraction based on VMD	902.50	96.1
The original signals	8873.96	98.6

Although the classification accuracy based on VMD feature extraction is 2.5% less than the direct classification, the training time is substantially shortened by 2.2 h. The reason for this result is that the proposed approach reduces the dimensionality of the data and the complexity of network calculation by VMD feature extraction, which shortens the classification time, but it is inevitable that some feature information is lost. In addition, Gaussian white noise is added to the original signals and show the classification results at a 15 dB, 20 dB and 25 dB signal-to-noise ratio (SNR) in Table 4.

Table 4. Classification accuracy under Gaussian white noise.

Methods	Classification Accuracy (%)		
	15 dB	20 dB	25 dB
Feature extraction based on VMD + CNN	83.52	89.33	93.21
The original signals + CNN	16.88	65.90	82.76

As reported in Table 4, the lower the SNR is, the better the robustness of the proposed approach. In particular, when the SNR is 15 dB, the original signals directly input into the CNN cannot be distinguished. However, the accuracy of the proposed approach is still as high as 83.52%. Thus, the results show that the VMD algorithm can accurately recognize failure under high-noise conditions.

To verify the effectiveness of the CNN model as a classifier, the classification results of four traditional classifiers (DT, KNN, BP and SVM) and the CNN are evaluated using the same training samples and test samples (as reported in Table 5).

Table 5. Comparison of different classifiers.

Classifiers	Accuracy (%)
Feature extraction based on VMD + DT	85.31
Feature extraction based on VMD + KNN	93.45
Feature extraction based on VMD + BP	79.10
Feature extraction based on VMD + SVM	80.74
Feature extraction based on VMD + CNN (ours)	96.10

In Table 5, the classification accuracy of DT is 85.31%, that of KNN is 93.45%, that of BP is 79.1%, and that of SVM is 80.74%. Compared with the other three traditional algorithms, KNN has better classification results, but its accuracy is still less than that of the proposed approach. This result occurs because in the proposed approach, the CNN performs quadratic feature extraction by convolution before providing output as a classifier; by obtaining additional depth information, it can accurately realize recognition and classification. On the other hand, the advantage of the CNN is that although it takes considerable time in data training, once the training model is completed, it maintains good predictability and adaptability to analyze the same type of data.

4. Discussion

The results reported in Section 3 show that the spectra of incipient cable failures and overcurrent disturbance signals consist of different frequency components. Therefore, the method of extracting fine time-frequency information by VMD separating different frequency components is reliable. Second, the classification results show that the feature extraction method based on mathematical statistics after VMD can effectively shorten the training time of the CNN.

The comparison results under noise conditions also show that the original signal is difficult to distinguish under the influence of high noise. Therefore, VMD improves the robustness of failure recognition by filtering high-frequency noise, resulting in more accurate classification results.

On the other hand, the comparison results of different classifiers show that the CNN, as a classifier with feature processing ability, can provide higher possibility of accurate classification.

Finally, it is worth noting here that the proposed approach is based only on simulated signals. Although all simulation samples were generated based on the actual interference waveform, more validation is needed if the approach is to be successfully applied to actual measurements.

5. Conclusions

To solve the problem that incipient cable failures are difficult to distinguish from overcurrent disturbances, an approach for the recognition and classification of incipient cable failures based on VMD and a CNN is proposed. First, the original signal is decomposed to obtain component signals containing more waveform features, and then the eigenvalues are calculated by analyzing the characteristics of the different signals; the results are used to construct an the input vector of the CNN. Finally, by training the CNN, recognition and classification of initial cable failure is achieved.

The approach makes full use of the noise immunity of VMD and the self-learning of the CNN, so achieves reliable and accurate classification. The effectiveness of the proposed approach is verified using different classifiers and high noise. The experimental results show that the approach has high precision and robustness.

Author Contributions: Data curation, W.Z.; Methodology, X.Y.; Software, J.D.

Funding: This research received no external funding.

Conflicts of Interest: The authors declare that they have no conflicts of interest to disclose.

References

1. Tang, Z.; Zhou, C.; Jiang, W. Analysis of Significant Factors on Cable Failure Using the Cox Proportional Hazard Model. *IEEE Trans. Power Deliv.* **2014**, *29*, 951–957. [CrossRef]
2. Bretas, A.S.; Herrera-Orozco, A.R.; Herrera-Orozco, C.A. Incipient fault location method for distribution networks with underground shielded cables: A system identification approach. *Int. Trans. Electr. Energy Syst.* **2017**, *27*, e2465. [CrossRef]
3. Zhang, W.; Xiao, X.; Zhou, K. Multi-Cycle Incipient Fault Detection and Location for Medium Voltage Underground Cable. *IEEE Trans. Power Deliv.* **2016**, *32*, 1450–1459. [CrossRef]
4. Charytoniuk, W.; Lee, W.J.; Chen, M.S. Arcing fault detection in underground distribution networks-feasibility study. *IEEE Trans. Ind. Appl.* **2000**, *36*, 1756–1761.
5. Pavlatos, C.; Vita, V.; Dimopoulos, A.C.; Ekonomou, L. Transmission lines' fault detection using syntactic pattern recognition. *Energy Syst.* **2018**, *10*, 299–320. [CrossRef]
6. Pavlatos, C.; Vita, V. Linguistic representation of power system signals. In *Electricity Distribution*; Energy Systems Series; Springer: Berlin/Heidelberg, Germany, 2016; pp. 285–295.
7. Xu, J.; Baese, U.M.; Huang, K. FPGA-based solution for real-time tracking of time-varying harmonics and power disturbances. *Int. J. Power Electron.* **2012**, *4*, 134. [CrossRef]
8. Romano, P.; Imburgia, A.; Ala, G. Partial Discharge Detection Using a Spherical Electromagnetic Sensor. *Sensors* **2019**, *19*, 1014. [CrossRef]
9. Nuruzzaman, A.; Boyraz, O.; Jalali, B. Time-stretched short-time Fourier transform. *IEEE Trans. Instrum. Meas.* **2006**, *55*, 598–602. [CrossRef]
10. Debnath, L. *Wavelet Transforms and Time-Frequency Signal Analysis. Wavelet Transforms and Time-Frequency Signal Analysis*; Birkhäuser: Boston, MA, USA, 2001.

11. Huang, N.E.; Shen, Z.; Long, S.R.; Wu, M.C.; Shih, H.H.; Zheng, Q.; Yen, N.-C.; Tung, C.C.; Liu, H.H. The Empirical Mode Decomposition and the Hilbert Spectrum for Nonlinear and Non-Stationary Time Series Analysis. *Proc. Math. Phys. Eng. Sci.* **1998**, *454*, 903–995. [CrossRef]

12. Cheng, H.; Chen, X.; Liu, F.; Wang, C. Series Arc Fault Detection and Implementation Based on the Short-time Fourier Transform. In Proceedings of the Asia-Pacific Power and Energy Engineering Conference, Chengdu, China, 28–31 March 2010.

13. Sidhu, T.S.; Xu, Z. Detection of Incipient Faults in Distribution Underground Cables. *IEEE Trans. Power Deliv.* **2010**, *25*, 1363–1371. [CrossRef]

14. Zhang, C.; Kang, X.N.; Ma, X.D. On-line Incipient Faults Detection in Underground Cables Based on Single-end Sheath Currents. In Proceedings of the Power & Energy Engineering Conference, Suzhou, China, 15–17 April 2016.

15. Gu, F.C.; Chang, H.C.; Chen, F.H. Application of the Hilbert-Huang transform with fractal feature enhancement on partial discharge recognition of power cable joints. *IET Sci. Meas. Technol.* **2012**, *6*, 440–448. [CrossRef]

16. Dragomiretskiy, K.; Zosso, D. Variational Mode Decomposition. *IEEE Trans. Signal Process.* **2014**, *62*, 531–544. [CrossRef]

17. Zhang, J.; He, J.; Long, J.; Yao, M.; Zhou, W. A New Denoising Method for UHF PD Signals Using Adaptive VMD and SSA-Based Shrinkage Method. *Sensors* **2019**, *19*, 1594. [CrossRef] [PubMed]

18. Achlerkar, P.D.; Samantaray, S.R.; Manikandan, M.S. Variational Mode Decomposition and Decision Tree Based Detection and Classification of Power Quality Disturbances in Grid-Connected Distributed Generation System. *IEEE Trans. Smart Grid* **2018**, *9*, 3122–3132. [CrossRef]

19. Voumvoulakis, E.M.; Gavoyiannis, A.E.; Hatziargyriou, N.D. Application of Machine Learning on Power System Dynamic Security Assessment. In Proceedings of the International Conference on Intelligent Systems Applications to Power Systems, Taiwan, China, 5–8 November 2007.

20. Tang, J.; Jin, M.; Zeng, F.; Zhou, S.; Zhang, X.; Yang, Y.; Ma, Y. Feature Selection for Partial Discharge Severity Assessment in Gas-Insulated Switchgear Based on Minimum Redundancy and Maximum Relevance. *Energies* **2017**, *10*, 1516. [CrossRef]

21. Mas'ud, A.; Albarracín, R.; Ardila-Rey, J.; Muhammad-Sukki, F.; Illias, H.; Bani, N.; Munir, A. Artificial Neural Network Application for Partial Discharge Recognition: Survey and Future Directions. *Energies* **2016**, *9*, 574. [CrossRef]

22. Mas'ud, A.; Ardila-Rey, J.; Albarracín, R.; Muhammad-Sukki, F.; Bani, N. Comparison of the Performance of Artificial Neural Networks and Fuzzy Logic for Recognizing Different Partial Discharge Sources. *Energies* **2017**, *10*, 1060. [CrossRef]

23. Parrado-Hernández, E.; Robles, G.; Ardila-Rey, J.; Martínez-Tarifa, J. Robust Condition Assessment of Electrical Equipment with One Class Support Vector Machines Based on the Measurement of Partial Discharges. *Energies* **2018**, *11*, 486. [CrossRef]

24. Shan, S. Decision Tree Learning. In *Machine Learning Models and Algorithms for Big Data Classification*; Springer: New York, NY, USA, 2016; pp. 237–269.

25. Zhang, S.; Wang, Y.; Liu, M. Data-based Line Trip Fault Prediction in Power Systems Using LSTM Networks and SVM. *IEEE Access* **2017**, *6*, 7675–7686. [CrossRef]

26. Cai, L.; Thornhill, N.; Kuenzel, S. Real-time Detection of Power System Disturbances Based on k-Nearest Neighbor Analysis. *IEEE Access* **2017**, *5*, 5631–5639. [CrossRef]

27. Maryam, M.N.; Flavio, V.; Taghi, M.K. Deep learning applications and challenges in big data analytics. *J. Big Data* **2015**, *2*, 1.

28. Wang, Y.; Liu, M.; Bao, Z. Deep learning neural network for power system fault diagnosis. In Proceedings of the 2016 35th Chinese Control Conference (CCC), Chengdu, China, 27–29 July 2016.

29. Hinton, G.E.; Osindero, S.; Teh, Y.W. A Fast Learning Algorithm for Deep Belief Nets. *Neural Comput.* **2014**, *18*, 1527–1554. [CrossRef]

30. Zhou, G.B.; Wu, J.; Zhang, C.L.; Zhou, Z.H. Minimal gated unit for recurrent neural networks. *Int. J. Autom. Comput.* **2016**, *13*, 226–234. [CrossRef]

31. Nguyen, M.; Nguyen, V.; Yun, S.; Kim, Y. Recurrent Neural Network for Partial Discharge Diagnosis in Gas-Insulated Switchgear. *Energies* **2018**, *11*, 1202. [CrossRef]

32. LeCun, Y.; Bottou, L.; Bengio, Y.; Haffner, P. Gradient-based learning applied to document recognition. *Proc. IEEE* **1998**, *6*, 2278–2324. [CrossRef]
33. Cheng, J.; Wang, P.S.; Gang, L.I. Recent advances in efficient computation of deep convolutional neural networks. *Front. Inf. Technol. Electron. Eng.* **2018**, *19*, 64–77. [CrossRef]
34. Bertsekas, D.P. *Constrained Optimization and Lagrange Multiplier Methods*; Academic Press: Cambridge, MA, USA, 1982; p. 104.

energies

MDPI

Article

Self-Shattering Defect Detection of Glass Insulators Based on Spatial Features

Haiyan Cheng [1], Yongjie Zhai [1], Rui Chen [1,*], Di Wang [2], Ze Dong [1] and Yutao Wang [1]

[1] Hebei Engineering Research Center of Simulation and Optimized Control for Power Generation, North China Electric Power University, Baoding 071003, China; chenghaiyan78@126.com (H.C.); zhaiyongjie@ncepu.edu.cn (Y.Z.); dongze33@126.com (Z.D.); 18730265696@163.com (Y.W.)

[2] Power China Guizhou Electric Power Design & Research Institute Co., LTD, Guiyang 550002, China; ncepuwangdi@163.com

* Correspondence: cr_chenrui@foxmail.com; Tel.: +86-136-6379-3681

Received: 21 January 2019; Accepted: 4 February 2019; Published: 10 February 2019

Abstract: During an automatic power transmission line inspection, a large number of images are collected by unmanned aerial vehicles (UAVs) to detect existing defects in transmission line components, especially insulators. However, with twin insulator strings in the inspection images, when the umbrella skirts of the rear string are obstructed by the front string, defect detection becomes difficult. To solve this problem, we propose a method to detect self-shattering defects of insulators based on spatial features contained in images. Firstly, the images are segmented according to the particular color features of glass insulators, and the main axes of insulator strings in the images are adjusted to the horizontal direction. Then, the connected regions of insulators in the images are marked. After that, the vertical lengths of the regions, the number of insulator pixels in the regions, as well as the horizontal distances between two adjacent connected regions are selected as spatial features, based on which defect discriminants are formulated. Finally, experiments are performed using the proposed formula to detect self-shattering defects in the insulators, using the spatial distribution of the connected regions to locate the defects. The experiment results indicate that the proposed method has good detection accuracy and localization precision.

Keywords: defect detection; glass insulator; localization; self-shattering; spatial features

1. Introduction

In recent years, the difficulty of conducting manual inspections on transmission lines has been increasing with the enlargement of transmission lines. In order to reduce the workload of inspection work and improve inspection efficiency, unmanned aerial vehicles (UAVs) are used for transmission line inspections [1–3]. During UAV inspections, a number of images are collected to detect the defects in transmission lines. The use of image detection technology greatly improves the efficiency and has become a research hotspot in current smart grids [4–7]. Insulators, of which the main roles are electrical insulation and line support, are very important components of transmission lines. Since glass insulators have the characteristics of zero value self-shattering, when an insulator is subjected to changes in terms of cold and heat, its compression stresses the surface and opposing tensile internal stresses become greater, which can cause breakdown easily. As a result, the insulation value at both ends of the insulator string becomes zero and the insulation function is lost. Therefore, zero value detection of insulators is required. When the insulation value is zero, a glass insulator will shatter, and it is easy to be found with visual detection. Additionally, because they are widely used in 500 kV high voltage transmission lines, detection of self-shattering defects in insulators has become a significant issue.

Therefore, many methods are suggested to detect self-shattering defects in glass insulators. They are as follows:

(1) Detection method based on contour information. Some researchers [8–11] firstly identified the contours of insulators, and then determined whether there were self-shattering defects according to the relative distances between adjacent insulators. Nevertheless, the insulators will sometimes be obscured because of unsuitable shooting angles of cameras. In addition, the backgrounds of the aerial images are complex and changeable. Therefore, the contour extraction of insulators will be influenced and the accuracy of the above detection method is affected.

(2) Detection method based on texture features. Zhang [12] and Wang [13] calculated the texture values of insulator images by block, and then analyzed whether the insulator was missing by comparing texture values. However, the calculation amounts of this method are very large. Zhang [14] obtained the directions of insulator strings by detecting parallel line segments. Then, the insulator string was divided into blocks according to the direction and the distance between the slices. After that the existence of self-shattering defects were judged by the similarity of texture features between the blocks. However, when the insulator pieces were obstructed or the shooting distance was large, the line features of the insulator pieces were very vague, which easily led to false detection.

(3) Detection method based on histogram matching features. Lin [15] diagnosed glass insulator faults based on a histogram matching criterion and determined faultless insulators within a certain range. However, the method could not indicate the specific locations of the fault insulators. Shi [16] used a sliding window method to match the gray histogram of the insulators and the templates, and inferred and located insulator defects according to histogram distances. However, it was easily influenced by the detection environment and the selection of templates.

(4) Detection method based on region features. Wang [17] segmented insulators in a lab color space and calculated the proportion of insulator pixels in each insulator region to determine the drop-off fault. Jiang [18] divided the insulator area into individual pieces, and judged the fault by sensing the distances between the gravity centers of adjacent insulator pieces. Nevertheless, when the insulator string was obstructed and the two insulator strings could not be separated completely, the detection accuracies of the two methods were influenced. Zhai [19] segmented insulator images in the RGB color space, and performed an adaptive morphology process on the images according to the area ratio of the insulator region. After that, the fault point was located. However, when the missing pieces were at the ends of the insulator string, they may have gone undetected.

The above-mentioned methods are mainly aimed at self-shattering defect detection of unobstructed glass insulators and have achieved good detection results within a certain range. However, in the process of transmission line inspections, insulators in the aerial images are often obstructed and also connected to each other because of the influence of the shooting distance and angle. Thus, most of the above methods show difficulty in obtaining the expected results.

Aiming at the above deficiencies and inspired by the literature [20], in this paper, we propose a self-shattering defect detection method for obstructed or unobstructed twin insulator string images based on the obvious spatial features of the insulator regions. The overall framework of this method is shown in Figure 1. The main technical contributions made in this work include: (1) For images of twin glass insulator strings, the self-shattering defect can be detected and the defect position can be located accurately, regardless of whether the insulators are obstructed or not; and (2) robustness and real-time performance is evaluated, and they may meet the requirements of strong robustness and high real-time performance demands of power line inspection. The remainder of this paper is organized as follows: In Section 2, the glass insulator image compound binarization processing are introduced. Then, in Section 3, the new detection and localization method of insulator self-shattering defects is described in detail. After that, in Section 4, experiments to verify the performance of the proposed method are explained. Finally, the conclusions are drawn in Section 5.

```
┌──────────────────────────┐        ┌──────────────────────────┐
│  Insulator aerial images │───────▶│      Tilt correction     │
└──────────────────────────┘        └──────────────────────────┘
            │                                     │
            ▼                                     ▼
┌──────────────────────────┐        ┌──────────────────────────┐
│    Image segmentation    │        │  Mark connected regions  │
└──────────────────────────┘        └──────────────────────────┘
            │                                     │
            ▼                                     ▼
┌──────────────────────────┐        ┌──────────────────────────┐
│  Morphological processing│        │ Calculate spatial features│
└──────────────────────────┘        └──────────────────────────┘
            │                                     │
            ▼                                     ▼
┌──────────────────────────┐        ┌──────────────────────────┐
│ Remove small area pseudo │        │ Detect self-explosion defect│
│         objects          │        └──────────────────────────┘
└──────────────────────────┘                     │
            │                                     ▼
            ▼                        ┌──────────────────────────┐
┌──────────────────────────┐        │ Locate self-explosion defect│
│    Hough line dection    │────────└──────────────────────────┘
└──────────────────────────┘
```

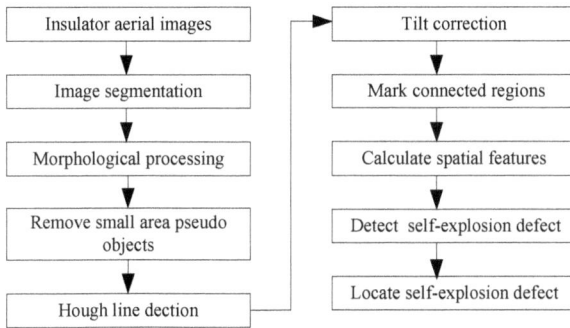

Figure 1. Flow chart showing the insulator self-shattering defect detection method.

2. Insulator Images Compound Binarization

The process of compound binarization mainly includes image segmentation, removal of pseudo targets and tilt correction.

2.1. Insulator Images Segmentation

Color is an intuitive feature of objects. Glass insulators without RTV spraying have a special color. It is generally green and very different to most background colors. Therefore, insulator images can be segmented according to their color features.

The authors of References [17,21] segmented insulator images using single thresholds in Lab space and HSI space, respectively. Because of the complex background in insulator images, it is difficult to achieve a good segmentation effect with a single threshold.

Inspection images collected by UAVs are usually RGB space images. In this paper, the R, G and B components of 150 glass insulator images were sampled, respectively, and their spatial distribution features were statistically analyzed. The color distribution rules of the glass insulators were as follows:

$$\left\{ \begin{array}{l} 65 \leq R \leq 175 \\ 115 \leq B \leq 180 \\ 30 \leq G - R \leq 65 \end{array} \right. \tag{1}$$

where R, G and B are the component values of red, green and blue, respectively.

The threshold values of the R, G and B components are set by Equation (1). Thus, the glass insulators can be easily separated from the complex background. Figure 2 shows the segmentation results of aerial insulator images according to Equation (1). Although the backgrounds of the two images were both very complex, good segmentation results were obtained regardless of whether the insulators were obstructed or not.

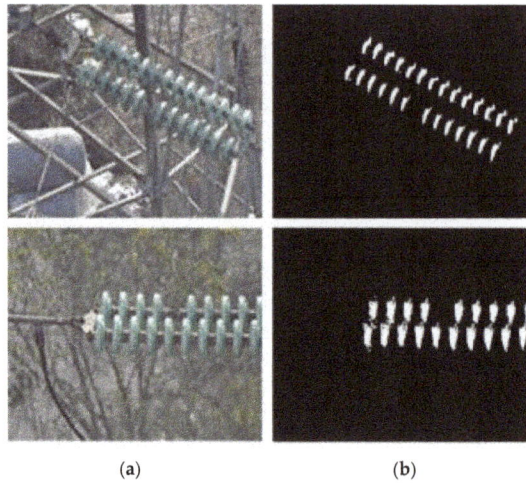

Figure 2. Results of segmentation: (**a**) Original aerial images; (**b**) Segmentation results.

2.2. Pseudo Objects Removal and Tilt Correction

After segmentation, the insulators images were separated from the background. However, there still existed some narrow breaks and small burrs in the images. Thus, the morphological closed operation and median filter were used to process the images [22]. After that, the small area pseudo objects were removed. Figure 3 shows the processing results of removal of the pseudo objects.

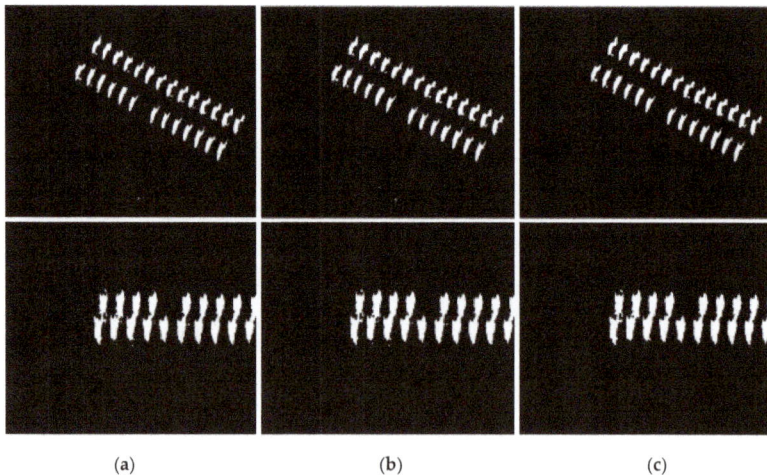

Figure 3. Pseudo objects removing process: (**a**) Morphological processing; (**b**) Median filter; (**c**) Pseudo objects removal.

Due to the influence of the shooting angle, insulators present different angles in the image. Thus, the image needs to be corrected to locate the main axis of the insulator string in the horizontal direction. Thus, it can make preparation for the spatial feature descriptions of the insulator regions. The direction of the longest line segment is the direction of the main axis of the insulator string in the image. Thus, the inclination angle of the main axis can be obtained using the Hough transform. According to the angle, the insulator string axis can be rotated in the horizontal direction. Figure 4 shows the processing results of tilt correction.

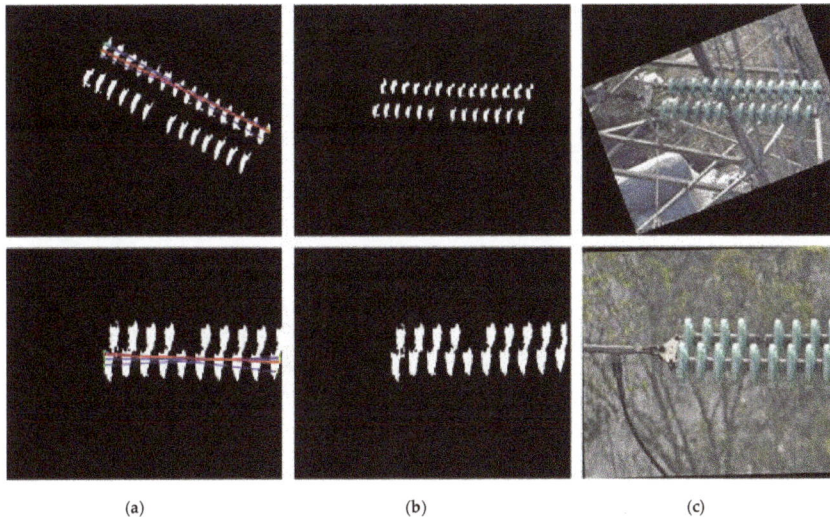

Figure 4. Tilt correction: (**a**) Hough transformation detection of lines; (**b**) Tilt correction; (**c**) Correction of original image.

It can be seen from Figures 3 and 4 that effective binarization images were obtained after image compound binarization. The contours of the insulators may be incomplete, but the spatial distribution of the insulator regions can be expressed regardless of if the insulators are obstructed or not.

3. Self-Shattering Defect Detection of Insulators

3.1. Spatial Features Description of Insulators

An insulator string is composed of a main shaft and a certain number of insulators. The insulators are perpendicular to the main shaft. The insulators on the same string have the same shape, size and color, and the distances between the adjacent umbrella skirts of the insulators are equal. All these characteristics indicate that insulators have consistent spatial features.

In order to represent the spatial features, the connected regions of the binary images are marked. The connected regions of normal insulators have three obvious features: (1) the vertical lengths of each connected region are close; (2) the number of insulator pixels in the region varies little; and (3) the horizontal distances between adjacent connected regions are very similar. In addition, the numerical distributions of the above three values are relatively uniform, as shown in Figure 5a. In Figure 5, the first row contains inspection images, corresponding connected regions of insulators are given in the second row, and the vertical length, the number of insulator pixels and the horizontal distance are in the third to the fifth rows, respectively.

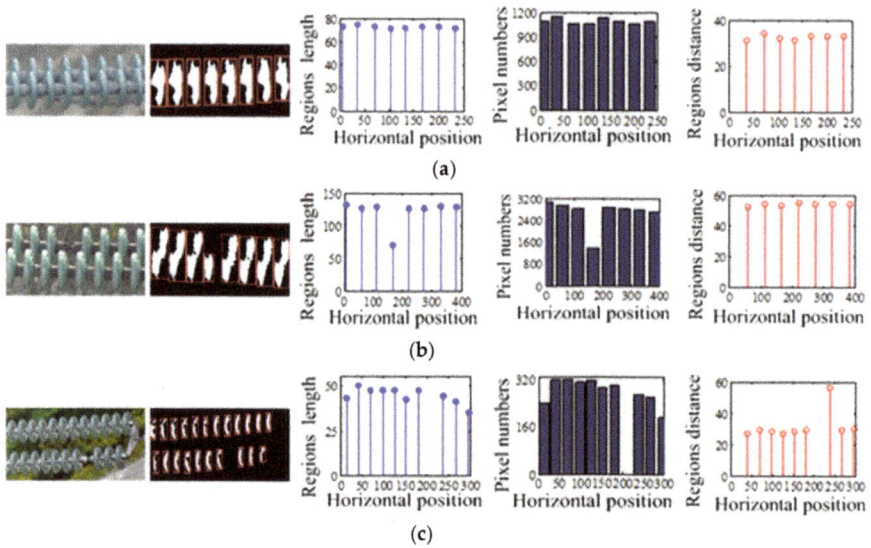

Figure 5. Spatial features of connected regions: (**a**) Normal; (**b**) Self-shattering with obstruction; (**c**) Self-shattering without obstruction.

When a self-shattering defect occurs, the spatial consistency of the insulator regions is destroyed, and spatial features change significantly. There are generally two cases of spatial locations of the twin insulator strings in the images. One is that the front string obstructs the back one, and the twin strings cannot be divided into individual strings in the image. Another situation is that the two insulator strings are not obstructed by each other, thus they can be separated into two strings. In the first situation, when the self-shattering defect occurs, the vertical lengths and the numbers of insulator pixels of the connected region are obviously smaller than those in the normal regions. Additionally, the numerical values are abruptly changed, as shown in Figure 5b. In the second situation, two insulator strings can be treated separately. When the self-shattering defect occurs, the horizontal distance between the connected regions on both sides of the defective insulator is obviously greater than those of other adjacent regions, and there is a clear jump, as shown in Figure 5c (Figure 5c presents the spatial features of the insulator string on which self-shattering occurs).

As can be seen from Figure 5, the spatial features of the self-shattering insulators are very different from the normal insulators. Therefore, the self-shattering defect can be discriminated according to the vertical lengths of the connected region, the number of insulator pixels in the region, and the horizontal distances between the adjacent regions.

3.2. Self-Shattering Defect Detection

3.2.1. Spatial Features Determination of Insulator Connected Regions

Following the above image processing, the main axis of the insulator string has been tilted in the horizontal direction. Then, the connected regions of the insulators are located using the 8-adjacency connected region method. For convenience of description, the connected region of each insulator is described by a four-dimensional vector (x, y, w, h). Where x and y correspond to the start point coordinates of the connected region on the x-axis and y-axis with the origin at the upper left corner, and w, h represent the horizontal width and vertical length of the region, respectively.

Based on analysis of the insulator spatial features, three spatial features, which characterize the connected regions of the insulators, are constructed as follows:

(1) Vertical length h_i: The vertical length of the connected region of the $i - th$ insulator.
(2) Number of insulator pixels n_i: The area of the connected region of the $i - th$ insulator.
(3) Horizontal distance between two adjacent connected regions D_i:

$$D_i = |x_{i+1} - x_i|; i = 1, 2, \cdots, k - 1 \tag{2}$$

where k is the number of connected regions, x_i is the horizontal coordinate of the starting point of the $i - th$ connected region.

3.2.2. Self-Shattering Defect Discriminant Definition

After determining the three spatial features of the insulators connected regions, self-shattering defect discrimination can be performed. Corresponding to the former two insulator string location situations, through statistical analysis, the discriminant$_1$ and discriminant$_2$ are summarized respectively in Equations (3) and (4):

$$\text{discriminant}_1 : \begin{cases} h_i \leq 0.7\overline{h} \\ n_i \leq 0.58\overline{n} \end{cases}; i = 1, 2, \cdots, k \tag{3}$$

where $\overline{h}$ and $\overline{n}$ are the average of h and n, respectively.

$$\text{discriminant}_2 : D_i > 1.7\overline{D}; i = 1, 2, \cdots, k - 1 \tag{4}$$

where $\overline{D}$ is the average of D.

When discriminant$_1$ or discriminant$_2$ is satisfied, it can be concluded that insulator self-shattering has occurred.

3.2.3. Self-Shattering Defect Localization

When the spatial features of the twin insulator strings satisfy discriminant$_1$, the latter insulator string is obstructed, and self-shattering may occur in either string. Thus, the defect location can be obtained by Equations (5)–(8).

$$x_{fault} = x_i \tag{5}$$

$$y_{fault} = \begin{cases} 0.5(y_{i-1} + y_{i+1}); y_i \gg \overline{y} \\ y_i + h_i; y_i \approx \overline{y} \end{cases} \tag{6}$$

$$w_{fault} = 0.5(w_{i-1} + w_{i+1}) \tag{7}$$

$$h_{fault} = 0.5(h_{i-1} + h_{i+1}) - h_i \tag{8}$$

where $\overline{y}$ is the average of y. For special cases, when self-shattering occurs at both ends of an insulator string, Equation (9) is used:

$$y_{i+1} = y_{i-1}, w_{i+1} = w_{i-1}, h_{i+1} = h_{i-1} \tag{9}$$

When the spatial features of the twin insulator strings satisfy discriminant$_2$, the insulators are not obstructed by each other. Thus, Equations (10)–(13) are adopted to calculate the defect location coordinates:

$$x_{fault} = x_i + 0.5D_i \tag{10}$$

$$y_{fault} = 0.5(y_i + y_{i+1}) \tag{11}$$

$$w_{fault} = 0.5(w_i + w_{i+1}) \tag{12}$$

$$h_{fault} = 0.5(h_i + h_{i+1}) \tag{13}$$

When there is no overlap between the twin insulator strings, they will be processed separately. In this case, if self-shattering occurs at both ends of one insulator string, the self-shattering defect can be judged based on the difference between the connected region numbers of the two insulator strings.

Figure 6 shows a flow chart of the self-shattering defect detection process after image compound binarization.

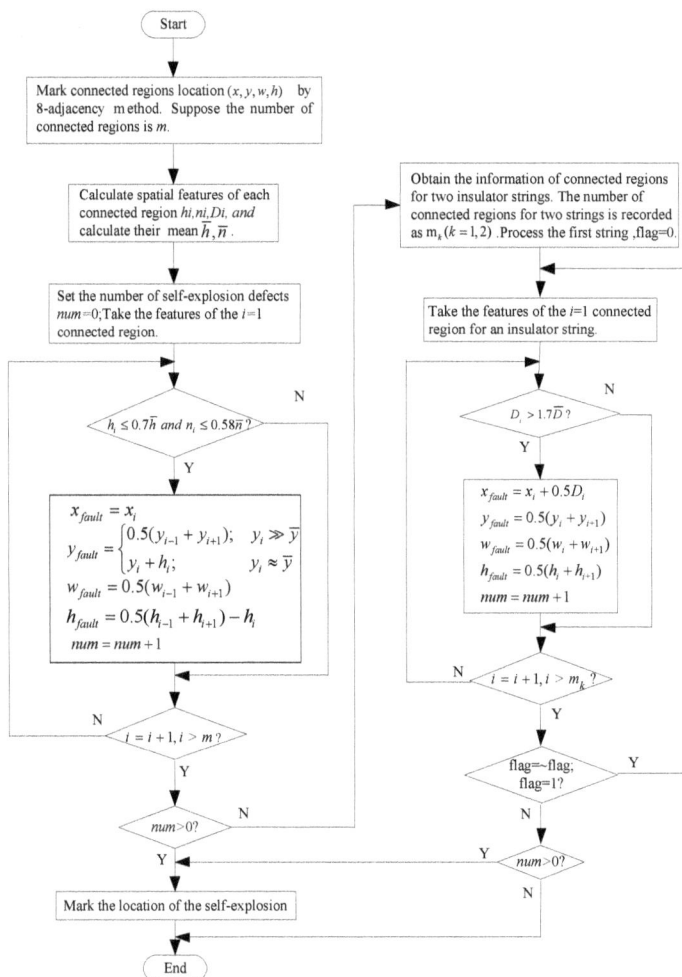

Figure 6. Flow chart of the proposed detection method.

When the back insulators are obstructed, the main pseudo code of the self-shattering detection is as follows (Algorithm 1):

Algorithm 1. Self-shattering detection process with obstructed.

Input: binarization image bwsrc
Output: Self-shattering location
1: $[B, \text{num}] = bwlabel(bwsrc, 8);$
2: $box = regionprops(B,' BoundingBox,' Area');$
3: $areas = [box.Area];$
4: $rects = cat(1, box.BoundingBox);$
5: $avgarea = mean(areas(:));$
6: $[m, n] = size(rects);$
7: $x = rects(:, 1); y = rects(:, 2);$
8: $w = rects(:, 3); h = rects(:, 4);$
9: $avgy = mean(y); avgh = mean(h);$
10: $hnorm = 0.7 * avgh; anorm = 0.58 * avgarea;$
11: $num = 0;$
12: **for** $j = 1\, to\, m$ **do**
13: **if** $h(j) <= hnorm \,\&areas(j) <= anorm$
14:    $num = num + 1;$
15:    $xx(num) = x(j);$
16:    $ww(num) = 0.5 * (w(j-1) + w(j+1));$
17:    $hh(num) = 0.5 * (h(j-1) + h(j+1)) - h(j);$
18: **if** $abs(y(j) - avgy) <= 10$
19:    $yy(num) = y(j) + h(j);$
20:      **else** $yy(num) = 0.5 * (y(j-1) + y(j+1));$
21: **end if**
22: **end if**
23: **end for**
24: **for** $k = 1\, to\, num$ **do**
25: $rec\tan gle('position', [xx(k), yy(k), ww(k), hh(k)]);$
26: **end for**

The self-shattering location detected by the proposed method is illustrated in Figure 7.

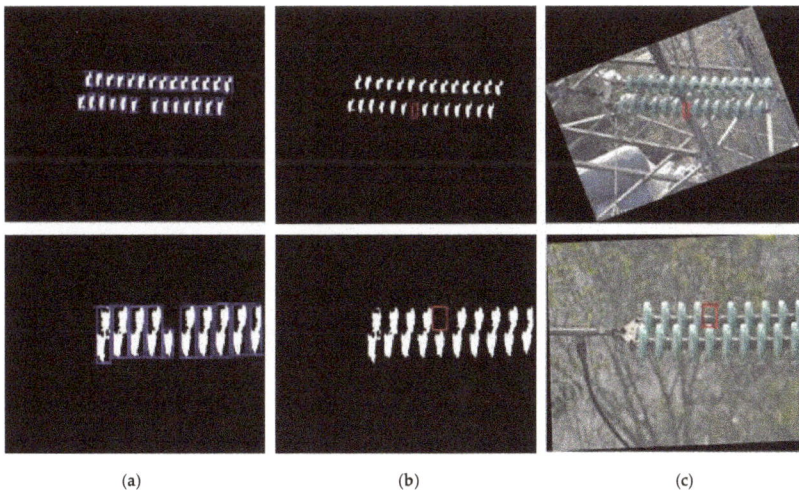

(a) (b) (c)

Figure 7. Results of localization: (**a**) Connected regions of insulators; (**b**) Self-shattering location; (**c**) Localization of original image.

4. Experimental Results and Analysis

4.1. Self-Shattering Detection and Localization Results Analysis

Experiments were conducted to evaluate the proposed method with real inspection images collected by UAVs. The experiment environments included the Windows 7 operating system with a CPU main frequency of 3.40 GHz, 4.00 GB RAM, and MATLAB 2015b. Figure 8 presents a part of the experimental results. The results show that the proposed method efficiently detected and located the insulator self-shattering defects.

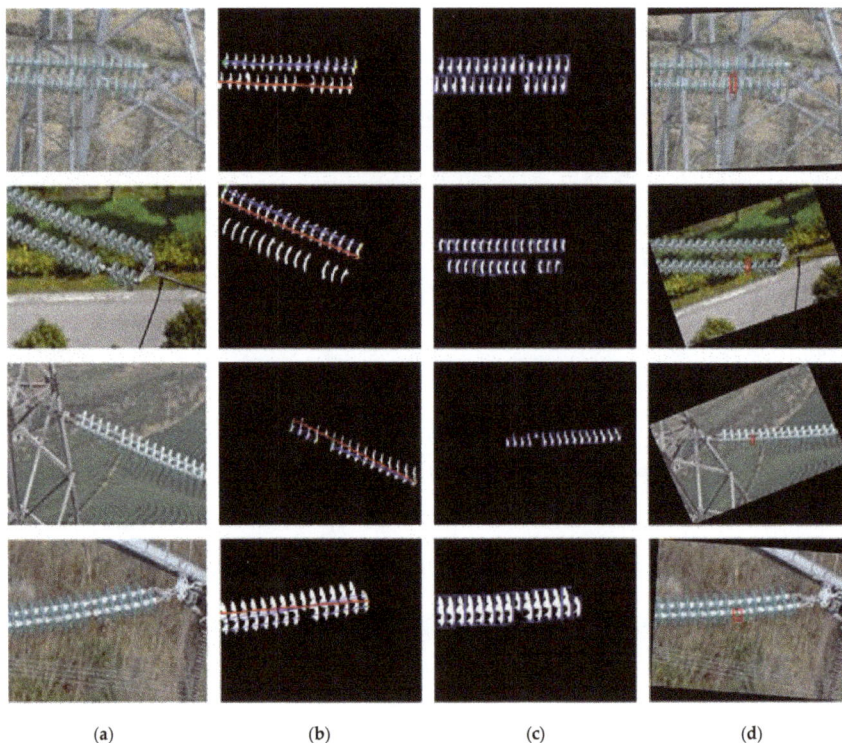

| (a) | (b) | (c) | (d) |

Figure 8. Test results: (a) Original image; (b) Detected lines; (c) Connected regions; (d) Defect localization.

4.2. Robustness Analysis

In this section, the robustness of the proposed method is analyzed through comparison with existing insulator self-shattering defect detection methods.

4.2.1. Detection Effect under Different Obstructed Conditions

The detection effect of the proposed method was compared with the methods in the literature [17,19,23] for inspection images with different obstruction conditions. The results are shown in Figures 9 and 10. It can be seen that the proposed method accurately located the self-shattering defects of the insulators in both conditions, and had strong robustness.

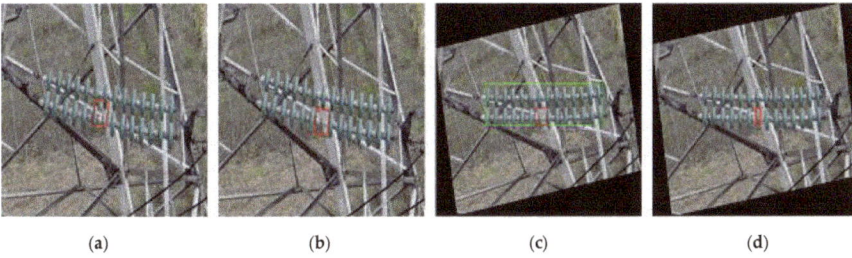

Figure 9. Comparison for insulators under the individual condition: (**a**) Literature [17] method; (**b**) Literature [19] method; (**c**) Literature [23] method; (**d**) Proposed method.

Figure 10. Comparison for insulators under the obstructed condition: (**a**) Literature [17] method; (**b**) Literature [19] method; (**c**) Literature [23] method; (**d**) Proposed method.

4.2.2. Detection Effects of Images with Different Background Complexities

The difficulty of self-shattering defect detection was different with different background complexities. The inspection images with different background complexities were used to further evaluate the performance of the proposed method. Experimental results are shown in Figure 11. It is demonstrated that the proposed method accurately located the self-shattering defects in inspection images with simple backgrounds and complex backgrounds.

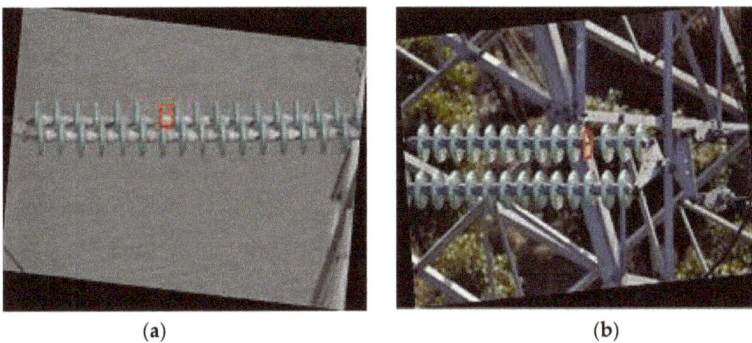

Figure 11. Results under different background complexities: (**a**) Simple background; (**b**) Complex background.

4.2.3. Self-Shattering Occurs at Both Ends of the Insulator String

The methods proposed in Reference [17,19] were not applicable for the situation where self-shattering occurs at both ends of one insulator string. However, the experimental results of the proposed method and the method used in Reference [23] are illustrated in Figure 12. It can be observed that literature [23] method got false self-shattering location, and the proposed method detected the self-shattering defect at one end of the insulator string, whereas it misjudged the visual vacancy on the other end, which led to error detection. From an operation and maintenance point of view, missed detection may lead to missing a fault, which can cause safety risks, while false detection may increase the workload of the secondary judgment. When the missed detection and the false detection cannot be completely avoided, in contrast, the choice is false detection rather than missed detection to improve safety. Therefore, the proposed method is more suitable for practical needs.

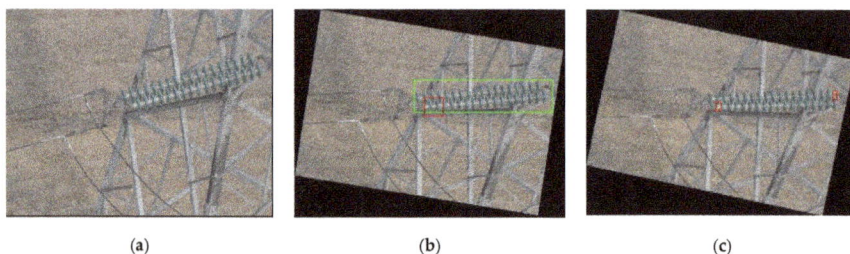

(a) (b) (c)

Figure 12. Detection result of self-shattering located at the end of insulator string: (a) Original image; (b) Literature [23] method; (c) Proposed method.

According to the above analysis, we conclude that the proposed method accurately detected and located the self-shattering defect of insulators in inspection images with different obstructed conditions, different self-shattering positions and different background complexities, which means that the proposed method had strong robustness.

4.3. Real-Time Performance and Precision

In order to analyze the real-time performance and detection precision of the proposed method, 67 inspection images of insulators with self-shattering defects were detected. The average time consumption, precision and missing rate were used as evaluation indexes. They are defined in Equations (14) and (15):

$$\text{precision} = TP/(TP + FP) \tag{14}$$

$$\text{missing rate} = FN/(TP + FN) \tag{15}$$

where TP represents the number of correctly located self-shattering insulators, $(TP + FP)$ is the total number of located self-shattering insulators, FN represents the number of undetected self-shattering insulators, and $(TP + FN)$ is the total number of self-shattering insulators.

The performances of the proposed method were compared with the methods in References [17,19,23], as shown in Table 1.

Table 1. Comparison of performances of proposed method with methods in References [17,19,23].

Method	Precision (%)	Missing Rate (%)	Average Time Consumption (s)
literature [17]	63.8	15.2	1.95
literature [19]	91.04	8.58	0.52
literature [23]	91.79	7.86	0.68
Proposed Algorithm	92.54	7.13	0.58

It was observed that the proposed method in this paper was superior to the method in References [17,23], which reduced the average time consumption and the missing rate of detection, and improved detection precision. Furthermore, compared with the method proposed in Reference [19], although our method took a slightly longer time, it had a higher precision and lower missing rate, which is more in line with actual needs.

5. Conclusions

In this paper, a detection and localization method for self-shattering defects in twin glass insulators, based on spatial features of connected regions, was proposed. In our method, firstly, the insulators in the inspection images were segmented from complex backgrounds based on color features, then insulator connected regions were marked, and finally insulators with defects could be detected and located based on spatial features. The performance of the proposed method was evaluated using inspection images under different conditions. Experimental results confirm that this method is simple, effective, and has a wide application range as well as good real-time performance.

In addition, when insulator regions are severely obstructed, an independent connected region based on the insulator umbrella skirt cannot be obtained, which results in poor performance of the method. Thus, further research based on spatial features is needed for the localization and detection of multiple and simultaneous bunch-drop faults in insulators.

Author Contributions: H.C., Y.Z., and Z.D. designed the system modeling and algorithm; H.C. and R.C. performed the experiments; H.C. and D.W. analyzed the experimental results; H.C., Y.Z., R.C. and Y.W. wrote this paper.

Funding: This study is supported by National Natural Science of Foundation of China (61773160, 61871182), and National Natural Science Foundation of Hebei Province (F2017502016).

References

1. Peng, X.; Qian, J.; Wang, K. Multi-sensor full-automatic inspection system for large unmanned helicopter and its application in 500 kv lines. *Guangdong Electr. Power* **2016**, *29*, 8–15.
2. Wang, M.; Du, Y.; Zhang, Z. Research on UAV aided inspection and image recognition of insulator defects. *J. Electron. Meas. Instrum.* **2015**, *29*, 1862–1869.
3. Guan, Z.; Wang, X.; Bian, X.; Wang, L.; Jia, Z. Analysis of causes of outdoor insulators damages on HV and UHV transmission lines in China. In Proceedings of the IEEE Electrical Insulation Conference, Philadelphia, PA, USA, 8–11 June 2014; pp. 227–230. [CrossRef]
4. Liao, S.; An, J. A robust insulator detection algorithm based on local features and spatial orders for aerial images. *IEEE Geosci. Remote Sens. Lett.* **2015**, *12*, 963–967. [CrossRef]
5. Cui, J.; Cao, Y.; Wang, W. Application of an improved algorithm based on watershed combined with krawtchouk invariant moment in inspection image processing of substations. *Proc. CSEE* **2015**, *35*, 1329–1335. [CrossRef]
6. Zhao, Z.; Xu, G.; Qi, Y.; Liu, N.; Zhang, T. Multi-patch deep features for power line insulator status classification from aerial images. In Proceedings of the International Joint Conference on Neural Network (IJCNN), Vancouver, BC, Canada, 24–29 July 2016; pp. 3187–3194. [CrossRef]
7. Yang, L.; Jiang, X.; Hao, Y.; Li, L.; Li, H.; Li, R.; Luo, B. Recognition of natural ice types on in-service glass insulators based on texture feature descriptor. *IEEE Trans. Dielectr. Electr. Insulation* **2017**, *24*, 535–541. [CrossRef]
8. Jiang, H.; Jin, L.; Yan, S. Recognition and fault diagnosis of insulator string in aerial images. *J. Mech. Electron. Eng.* **2015**, *32*, 274–278.
9. Zhang, S.; Yang, Z.; Huang, X.; Wu, H.Q.; Gu, Y.Z. Defects detection and positioning for glass insulator from aerial images. *J. Terahertz Sci. Electron. Inf. Technol.* **2013**, *11*, 609–613.
10. Shang, J.; Li, C.; Chen, L. Location and detection for self-explode insulator based on vision. *J. Electron. Meas. Instrum.* **2017**, *31*, 844–849.
11. Oberweger, M.; Wendel, A.; Bischof, H. Visual recognition and fault detection for power line insulators. In Proceedings of the 19th Computer Vision Winter Workshop, Krtiny, Czech Republic, 3–5 February 2014; pp. 1–8.

12. Zhang, X.; An, J.; Chen, F. A method of insulator fault detection from airborne images. In Proceedings of the 2nd WRI Global Congress on Intelligent Systems, Wuhan, China, 16–17 December 2010; pp. 200–203. [CrossRef]
13. Wang, W.; Wang, Y.; Han, J. Recognition and drop-off detection of insulator based on aerial image. In Proceedings of the 9th International Symposium on Computational Intelligence and Design, Hangzhou, China, 10–11 December 2016; pp. 162–167. [CrossRef]
14. Zhang, J.; Han, J.; Zhao, Y. Insulator recognition and defects detection based on shape perceptual. *J. Image Graphi.* **2014**, *19*, 1194–1201.
15. Lin, J.; Han, J.; Chen, F.; Xu, X.; Wang, Y. Defects detection of glass insulator based on color image. *Power Syst. Technol.* **2011**, *35*, 127–133.
16. Shi, L. The Method of Image Detection on Defect Insulator in Transmission Line. Master's Thesis, Department of Control Engineering, North China Electric Power University, Beijing, China, March 2013.
17. Wang, Y.; Yan, B. Vision based detection and location for cracked insulator. *Comput. Eng. Design* **2014**, *35*, 583–587.
18. Jiang, Y.; Han, J.; Ding, J. The identification and diagnosis of self-blast defects of glass insulators based on multi-feature fusion. *Electr. Power* **2017**, *50*, 52–58.
19. Zhai, Y.; Wang, D.; Zhang, M.; Wang, J.; Guo, F. Fault detection of insulator based on saliency and adaptive morphology. *Multimed. Tools Appl.* **2017**, *76*, 12051–12064. [CrossRef]
20. Zhai, Y.; Wang, D.; Zhao, Z.; Cheng, H. Insulator string location method based on spatial configuration consistency feature. *Proc. CSEE* **2017**, *37*, 1568–1577. [CrossRef]
21. Yao, C.; Jin, L.; Yan, S. Recognition of insulator string in power grid patrol images. *J. Syst. Simul.* **2012**, *24*, 1818–1822.
22. Jia, L.; Song, S.; Yao, L.; Li, H.; Zhang, Q.; Bai, Y.; Gui, Z. Image denoising via sparse representation over grouped dictionaries with adaptive atom size. *IEEE Access* **2017**, *5*, 22514–22529. [CrossRef]
23. Zhai, Y.; Chen, R.; Yang, Q.; Li, X.; Zhao, Z. Insulator fault detection based on spatial morphological features of aerial images. *IEEE Access* **2018**, *6*, 35316–35326. [CrossRef]

energies

MDPI

Article

Adaptive Nonlinear Model Predictive Control of the Combustion Efficiency under the NOx Emissions and Load Constraints

Zhenhao Tang *, Xiaoyan Wu and Shengxian Cao

School of Automation Engineering, Northeast Electric Power University, Jilin 132012, China;
wuxiaoyanneepu@hotmail.com (X.W.); caoshengxian@neepu.edu.cn (S.C.)
* Correspondence: tangzhenhao@neepu.edu.cn

Received: 19 March 2019; Accepted: 1 May 2019; Published: 8 May 2019

Abstract: A data-driven modeling method with feature selection capability is proposed for the combustion process of a station boiler under multi-working conditions to derive a nonlinear optimization model for the boiler combustion efficiency under various working conditions. In this approach, the principal component analysis method is employed to reconstruct new variables as the input of the predictive model, reduce the over-fitting of data and improve modeling accuracy. Then, a k-nearest neighbors algorithm is used to classify the samples to distinguish the data by the different operating conditions. Based on the classified data, a least square support vector machine optimized by the differential evolution algorithm is established. Based on the boiler key parameter model, the proposed model attempts to maximize the combustion efficiency under the boiler load constraints, the nitrogen oxide (NOx) emissions constraints and the boundary constraints. The experimental results based on the actual production data, as well as the comparative analysis demonstrate: (1) The predictive model can accurately predict the boiler key parameters and meet the demands of boiler combustion process control and optimization; (2) The model predictive control algorithm can effectively control the boiler combustion efficiency, the average errors of simulation are less than 5%. The proposed model predictive control method can improve the quality of production, reduce energy consumption, and lay the foundation for enterprises to achieve high efficiency and low emission.

Keywords: Combustion efficiency; NOx emissions constraints; boiler load constraints; least square support vector machine; differential evolution algorithm; model predictive control

1. Introduction

Large inertia and time delay of the coal-fired boiler burning process leads to a complicated model for a power station boiler. This makes its direct control difficult [1]. The combustion situation could be directly described with the boiler combustion efficiency. Thus, it could be considered as an important feature for boiler control and optimization. On the other hand, nitrogen oxides (NOx) generated from combustion emissions are known as significant environmental pollutants [2]. Thus, combustion optimization should be considered due to the economy and safety necessities. A high precision predictive model is required for combustion optimization.

Mechanism models, statistic models, and data-driven models are the three main models that have been proposed in the literature for the estimation of boiler combustion efficiency. A variety of mechanism models have been derived for the boiler combustion efficiency using the physical principles. For example, the superiority of a gray-box system identification method to the computational fluid dynamics (CFD) method for NOx emission modeling in a coal-fired power plant has been demonstrated [3]. An energy balancing-based model has been employed in Reference [4] to propose an effective method for the boiler combustion efficiency estimation. According to the experimental results presented in these papers, a good approximation could be achieved through the proposed models under certain constraints, however, their performance could be influenced by real operation parameters. This limits their application in practical production. Accordingly, employing these models in modern control methods is difficult. Several statistics-based models have been proposed for boiler combustion efficiency prediction in the literature [5–7]. In Reference [5], an autoregressive integrated moving average (ARIMA) model has been proposed, based on simulation data to predict the boiler combustion efficiency. The thermodynamic principles have been employed to develop a dynamic model for an ultra-supercritical coal-fired once-through boiler-turbine unit [6]. In Reference [7], the prediction accuracy of the boiler combustion efficiency has been verified through a partial least squares (PLS) model. Despite their simplicity and usefulness, these models are too sensitive to measurement errors. This makes them unsuitable for designing a model-based controller. Various data-driven models have been utilized in optimization-based controllers. Artificial neural networking (ANN) [8], back-propagation (BP) [9], and multilayer perceptron (MLP) neural networks [10] have been employed to predict the boiler combustion efficiency. Although the prediction accuracy of these models is satisfactory, over-fitting phenomena could have occurred in them. Moreover, a least square support vector machine (LSSVM) model has been extracted from the real data to obtain the optimal combustion parameters [11,12]. Here, the LSSVM was employed to construct the boiler load, the NOx emissions, and the boiler combustion efficiency models of the coal-fired boiler. Moreover, the differential evolution (DE) algorithm has been utilized for online optimization of the LSSVM parameters for different problems.

The experimental simulation data has been collected from the supervisor information system (SIS) or distribution control system (DCS). To improve the prediction accuracy and reduce the computation time, principal component analysis (PCA) has been selected to process the experimental data [13,14]. In addition, different operating conditions could be distinguished via the k-nearest neighbor (KNN) classifier. Accordingly, the modeling prediction accuracy could be improved.

Various predictive model-based optimization methods have been utilized to control the boiler combustion efficiency. In Reference [15], a new hybrid jump particle swarm optimization (PSO) algorithm has been proposed to tune the proportional integral (PI) controller that has been designed for the boiler-turbine unit control. Moreover, a genetic algorithm (GA) has been utilized to obtain the optimal controller parameters of the low NOx emissions in the presence of the invariable load [16]. The active disturbance rejection controller (ADRC) was proposed by Han, and it had a large stability margin and bandwidth. However, the method required the control torque disturbance to be set [17]. A new hybrid model-free control and virtual reference feedback tuning (MFC-VRFT) method was *f* proposed to solve the complex problem of traditional controllers. This method can solve the controller optimization problem well. However, multiple PI controller parameters must be considered [18]. To improve the boiler efficiency under the load constraint by finding the optimal parameters, the DE algorithm has been employed too [19–22]. Due to its highly robust performance, the model predictive control (MPC) has been employed in various control applications, such as civil engineering control [19,20], machinery manufacturing control [21] and steel industry control [22].

Improving the boiler combustion efficiency under NOx emissions and load constraints by using a nonlinear adaptive model predictive controller is the main goal of the current paper. A DE-based LSSVM algorithm and the feature selection approach are employed to extract appropriate models for the NOx emissions, the boiler load, and the boiler combustion efficiency for different operation conditions. This predictive model is employed to present an optimization model for improving boiler combustion efficiency under NOx emissions and boiler load constraints. Finally, a DE algorithm is utilized for solving the optimization model or equivalently finding the optimal control parameters.

This paper is organized as follows. The boiler combustion process is illustrated in Section 2. A nonlinear predictive model for the coal-fired boiler is extracted in Section 3 and its performance is validated through appropriate comparative models. An optimized output control scheme for the coal-fired boiler is presented in Section 4. The effectiveness of the proposed control structure is investigated through simulations. The concluding remarks are given in Section 5.

2. Background

A super-critical 600MW coal-fired boiler produced by the Harbin Boiler Plant Limited Liability Company is considered. The coal preparation system, the burner, the heating surface, and the air preheater are its main components. The boiler production process is shown in Figure 1. And the relation variables measurement points are marked in the figure. The coal in the coal hopper is fed into the coal mill via a belt conveyor, the pulverized coal and the primary air mixture via the milling system are fed into the combustion chamber. The steam generator combustion releases energy to produce saturated steam. Then, the superheated steam of a specific temperature is collected into main-pipe steam via the superheater, the superheated steam with a certain pressure enters the turbine's cylinder to drive the generator to generate electricity. Meanwhile, the flue gas is generated in the combustion process. In addition, to convert saturated steam into superheated steam, preheated air is preheated by boiler feed water and the air preheater via the economizer.

Flue gas is extracted from the boiler through the appropriate exhaust fans. The flue dust is separated from the electrostatic precipitate after exhausting flue gas into the atmosphere via an 80m high stack. The bottom ash (slag) and the flue dust are disposed of via the hydraulic system and transported through the pipelines to the landfill. When the furnace flame temperature rises above 900 °C in the boiler combustion process, the coal undergoes thermal decomposition and is further oxidized into NOx, accompanied by the reduction reaction of NOx. The producing and reducing NOx are not only related to the characteristics of coal, the state of NOx in coal, the distribution ratio of nitrogen in volatile and coke during nitrogen thermal decomposition of coal and their respective components, but is also related to the oxygen concentration and combustion temperature. The NOx produced by the coal combustion in the boiler is discharged from the chimney into the surrounding environment. It could be harmful to the atmosphere and the environment. Moreover, the boiler load leads to unsafe boiler running and a low fuel utilization ratio. Therefore, the operation variables should be controlled to achieve the optimal boiler combustion efficiency under the NOx emissions and boiler load constraints. It leads to an interesting research area.

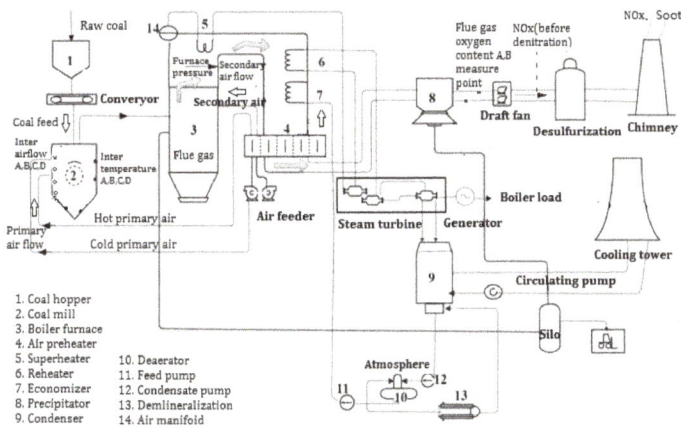

Figure 1. The description of coal-fired boiler production process.

3. Nonlinear Dynamic Prediction Model

The NOx emissions, the boiler load, and the boiler combustion efficiency could be described through nonlinear dynamic prediction models. These models are described in this section. The PCA algorithm is employed to reconstruct new variables as the predictive model inputs (see Section 3.1). This provides relevant information on the input parameters. The basic principles of the KNN classifier are introduced in Section 3.2. The prediction model is constructed in Section 3.3. The evaluation indices for model performance verification are defined in Section 3.4. Finally, the prediction performance of the proposed model is verified in Section 3.4.

3.1. Data Selection

The boiler efficiency, boiler load and NOx emission depend on various parameters, such as the coal feed, the main feed-water flow, the primary airflow, and the secondary airflow. The combustion mechanism and empirical analysis give eighteen boiler parameters that are gathered from the DCS. These parameters are defined in Table 1. The parameter name in the table is important related variables of the boiler efficiency (B), boiler load (L) and NOx emissions (N_x). Since the mentioned parameters have a different range, the data should be standardized. This improves the prediction accuracy. All the data are scaled according to the following equation:

$$S_n^k = \frac{X_n^k - \frac{1}{K}\sum_{k=1}^K X_k^n}{\sqrt{\frac{1}{K-1}\sum_{k=1}^K \left(X_k^n - \frac{1}{K}\sum_{k=1}^K X_k^n\right)^2}} \tag{1}$$

where S_k^n represents the n dimensional production parameters under data standardization, X_k^n represents the meta-data with the dimension n for the kth sample, and K is the number of the data set.

Table 1. The main boiler parameters.

Name		Nomenclature	Symbol	Unit	Classical
Input variable	Controllable variable	Coal feed	Fc	t/h	B/L/N_x
		Main feed-water flow	M	t/h	B/L/N_x
		Coal mill inlet airflow A	A_1	t/h	B/L/N_x
		Coal mill inlet airflow B	A_2	t/h	B/L/N_x
		Coal mill inlet airflow C	A_3	t/h	B/L/N_x
		Coal mill inlet airflow D	A_4	t/h	B/L/N_x
		Primary air flow	P_a	t/h	B/L/N_x
		Secondary air flow	S	t/h	B/L/N_x
	State variable	The flue gas oxygen content A	O_1	%	B/N_x
		The flue gas oxygen content B	O_2	%	B/N_x
		Coal mill inlet temperature A	T_1	°C	B
		Coal mill inlet temperature B	T_2	°C	B/N_x
		Coal mill inlet temperature C	T_3	°C	B
		Coal mill inlet temperature D	T_4	°C	B/N_x
		Furnace pressure	P	Pa	B
Output variable		NOx emissions in flue gas	N_x	mg/m^3	
		Boiler combustion efficiency	B	%	
		Boiler load	L	MW	

Due to the interaction between these parameters, the obtained prediction accuracy is low. Thus, the PCA method is employed for generating the required input for the prediction model and reducing the input dimension. The input variables are composed of controllable variables and state variables. $Fc, M, A_1, A_2, A_3, A_4, P_a, S$ are the most representative control variables in Table 1.

3.2. Classification of Operating Conditions

The actual production process runs under several operating conditions, and their parameters may change. Thus, a KNN algorithm is applied to separate different operating conditions. The K-value selection, distance measurement and the classification rule are three main stages in the KNN classification method. Firstly, the KNN algorithm selects the K value of the fixed sample data via cross-validation. Secondly, the distance measurement function is carried out for different operating conditions. The Euclidean distance describing with the following equation is selected for the distance measurement:

$$d_{1,2} = \sqrt{\sum_{l=1}^{n} (S_{1l} - S_{2l})^2}$$ (2)

where $d_{1,2}$ is the Euclidean distance between two actual production parameters with the dimension n, and S_{1l} and S_{2l} are the lth components of their corresponding standard deviations.

The test set is input and compared with a training set, the corresponding feature when the data and label in the training set are thus known. Corresponding features are determined by computing the Euclidean distance between the objects as a non-similarity index objects. Finally, the majority voting method is employed to classify the new sample data, and the corresponding model is utilized for the prediction. The greedy algorithm is employed to distinguish different operating conditions from the standard deviation distance of the clustering center. The sample data are divided into two groups. In this study, the boiler combustion efficiency, the boiler load, and the NOx emissions are considered as the operation conditions, respectively.

3.3. Construction of the Predictive Model

The LSSVM is adopted to construct the predictive model while the DE algorithm is utilized to find the optimal parameters of the LSSVM. The boiler load, NOx emissions and boiler combustion efficiency are divided into the H group data under the classified working conditions, then step 1 to 6 is executed, respectively. Each set of data represents a class of working conditions. The flowchart of the DE-based LSSVM (DELSSVM) algorithm is shown in Figure 2. The measurement models under corresponding working conditions are established by using the DE algorithm to the optimal LSSVM method. A description of the relationship between the input parameters and the boiler major parameters is given below:

Step 1: Divide data 1 (the NOx emissions samples set), data 2 (the boiler combustion efficiency samples set), and data 3 (the boiler load samples set) into training and test data, in accordance with the H groups operating conditions.

Step 2: Initialize the DE swarm randomly within the search space and adjust these parameters: Np as the population size, D as the population dimension, g as the iteration number, G as the maximum iterations, F and CR as the alternating and mutant probabilities, C and σ^2 as the regularization and kernel parameters ($Np = 500$, $D = 2$, $G = 1000$, $F = 0.7$, $CR = 0.9$ are considered in this paper).

Step 3: Construct the LSSVM model using each individual correspondent with the kernel parameters value of the LSSVM and the model data of the hth working conditions, the predictive model number of Np are obtained by training, radial based function (RBF) kernel function is used in the LSSVM model:

$$f(x) = \sum_{k=1}^{K} a_h K(x_k, x) + b \ \ k = 1, 2, \ldots, K$$ (3)

$$K(x_k, x) = \exp\left|\left(\frac{\|x_k - x\|^2}{2\sigma^2}\right)\right|$$ (4)

where $f(x)$ is the predictive value, a_h and b represent model parameters, x_k is the hth input parameter of the training sample, K is the number of training sets, $K(x_k, x)$ are the LSSVM kernel parameters, σ^2 is the width of the RBF function, and the DE algorithm is used to obtain σ^2.

Step 4: Calculate the prediction root mean square error ($\varepsilon(k)$) for every particle ($f(k)$) as:

$$\varepsilon(k) = \sqrt{\left(\sum_{k=1}^{N} |Y_k - f(x_k)|^2\right)/N}$$ (5)

where Y_k is the kth true value and $f(x_k)$ is the kth prediction value, and N is the number of data sets.

Step 5: If $g = G$, stop the algorithm and go to Step 6. Otherwise, update the particles according to Equations (6)–(8) and return to Step 3.

$$V_{j,g+1} = x_{j,g} + F(x_{best.g} - x_{j,g}) + F(x_{r1,g} - x_{r2,g})$$
$$g = 1, \ldots, G, j = 1, \ldots, N$$
(6)

$$U^i_{j,g+1} = \begin{cases} V^i_{j,g}, rand_i(j) \le CR, i = randn_j \\ x^i_{j,g}, rand_i(j) > CR, i \ne randn_j \end{cases}$$
$$i = 1, 2, \ldots, D, g = 1, \ldots, G, j = 1, \ldots, N$$
(7)

$$x_{j,g+1} = \begin{cases} U_{j,g}, f(U_{j,g}) < f(x_{j,g}) \\ x_{j,g}, f(U_{j,g}) \ge f(x_{j,g}) \end{cases}$$
$$g = 1, \ldots, G, j = 1, \ldots, N$$
(8)

where $V_{j,g+1}$ is the jth individual in the $(g + 1)$th iteration; $x_{best.g}$ is the best individual in iteration g; j, $r1$, $r2$ are different random integer numbers in the interval $[1, Np]$; $F\in [0, 2]$ is the alternating probability; $U^i_{j,g+1}$ is the jth individual in the $(g + 1)$th iteration and the dimension parameter of the ith individual; $rand$ () means a random number with a uniform distribution in the interval $[0, 1]$; $CR\in [0, 1]$ is the crossover probability and $f(*)$ is the individual fitness function value.

Step 6: Calculate the optimal penalty factor C and the kernel function σ^2 as well as the predictive model output.

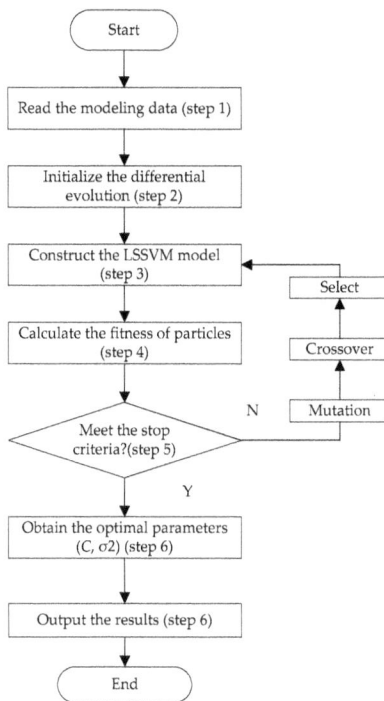

Figure 2. Flowchart of the differential evolution-based least square support vector machine (DELSSVM) algorithm.

The model input variables are composed of state variables and controllable variables given in Table 1. And the prediction models for the boiler combustion efficiency, the NOx emissions and boiler load for different operating conditions are described as:

$$Y_{B,h} = f(Fc, M, A_1, A_2, A_3, A_4, P_a, S, O_1, O_2, T_1, T_2, T_3, T_4, P, \sigma_B^2, w_B, C, e_{i,B}, a_{i,B}, b_B, B^*) \tag{9}$$

$$Y_{Nx,h} = f(Fc, M, A_1, A_2, A_3, A_4, P_a, S, O_1, O_2, T_1, T_2, \sigma_{N_x}^2, w_{Nx}, C_{Nx}, e_{i,Nx}, a_{i,Nx}, b_{Nx}, N_x^*) \tag{10}$$

$$Y_{L,h} = f(Fc, M, A_1, A_2, A_3, A_4, P_a, S, \sigma_L^2, w_L, C_L, e_{i,L}, a_{i,L}, b_L, L^*) \tag{11}$$

where $Y_{B,h}$, $Y_{Nx,h}$ and $Y_{L,h}$ are the predicted values for the boiler efficiency, the NOx emissions and the boiler load under the hth operating condition, respectively; Fc, M, A_1, A_2, A_3,A_4, P_a, S, O_1, O_2,T_1, T_2, T_3, T_4, P are the corresponding symbols for operating parameters given in Table 1; $\sigma_B{}^2$, $\sigma_{N_x}^2$ and σ_L^2 are the kernel parameters of boiler efficiency, NOx emissions and boiler load, respectively; w_B, w_{Nx} and w_L are the weight vectors of boiler efficiency, NOx emissions and boiler load, respectively; $e_{i,b}$, $e_{i,Nx}$ and $e_{i,L}$ are the loss functions of the boiler efficiency, NOx emissions and boiler load, respectively; $a_{i,B}$, $a_{i,Nx}$ and $a_{i,L}$ are the lagrange multipliers of boiler efficiency, NOx emissions and boiler load, respectively; b_B, b_{Nx} and b_L are the constants of boiler efficiency, NOx emissions and boiler load, respectively; N_x^*, B^* and L^* are the previous values for the boiler combustion efficiency, the NOx emissions and the boiler load (at previous sample time), respectively.

3.4. Model Validation and Analysis

3.4.1. Experiment Setup

The experimental data collected from the distribution control system (DCS) in the power plant are defined in Table 2. The positive balance method is adopted to calculate the boiler efficiency according to the relevant factors (input variables) of boiler efficiency. The input variables of boiler efficiency are shown in Table 1. The input variables of the predictive model are obtained according to Equations (9–11) under two operating conditions. All of these simulations are performed through the MATLAB R2014a environment on a PC with Intel Core TM i5 (2.50 GHz) processor, and 2.0 GB RAM, and Windows 10 32bit operating system.

Table 2. Dataset1.

Parameter	Number (Condition 1)			Number (Condition 2)		
	N_x	B	L	N_x	B	L
Train data (samples)	800	800	800	600	600	600
Test data (samples)	300	300	300	270	270	270
Sample Interval (min)	1	1	1	1	1	1
Input variable	13	16	9	13	16	9
Output variable	1	1	1	1	1	1

3.4.2. The Model Performance Evaluation Indicators

To compare the performance of the proposed model with other models, three evaluation indicators are defined in Table 3. These indicators could evaluate model prediction accuracy. These indicators include mean absolute error (MAE), mean absolute percentage error (MAPE), root mean square error (RMSE). These indicators can evaluate the average prediction ability of the model for each data point.

Table 3. The indicators for the model performance evaluation.

Metric	Definition	Equation
MAE	Mean absolute error	$MAE = (\sum_{k=1}^{K}|Y_k - \hat{Y}_k|)/K$
MAPE	Mean absolute percentage error	$MAPE = (\sum_{k=1}^{K}|Y_k - \hat{Y}_k|/Y_k)/K$
RMSE	Root mean square error	$RMSE = \sqrt{(\sum_{k=1}^{K}|Y_k - \hat{Y}_k|^2/(K-1))}$

3.4.3. Experimental Result Analysis

In this section, the approach DDMMF is proposed as the predictive model in this paper, the prediction performance of the proposed model is compared with the ARIMA model in Reference [5], the PLS model in Reference [7], and the MLP model in Reference [10]. In the first case, the predicted results of the proposed model for operating condition 1 are compared with the corresponding ones obtained from the MLP, ARIMA and PLS models. The predicted values for the boiler combustion efficiency, the NOx emissions, and the boiler load for different prediction models are shown in Figure 3, while their corresponding prediction errors are compared in Figure 4. The evaluation indicators obtained for these predicted values are compared for various prediction models. These comparative results are given in Table 4. The predicted values for operating condition 2 are compared in Figure 5, while their corresponding prediction errors are given in Figure 6. Table 5 compares the prediction evaluation indicators for operating condition 2.

For operating condition 1, the best fit to the real values was achieved by the DDMMF algorithm (see Figures 3 and 4). Moreover, Table 4 demonstrates that the smallest evaluation values, or equivalently, the best prediction capability could be obtained by the proposed method. The same results for operating condition 2 could be derived from Figures 5 and 6 and Table 5. In summary, the following quantitative results could be obtained from these simulations.

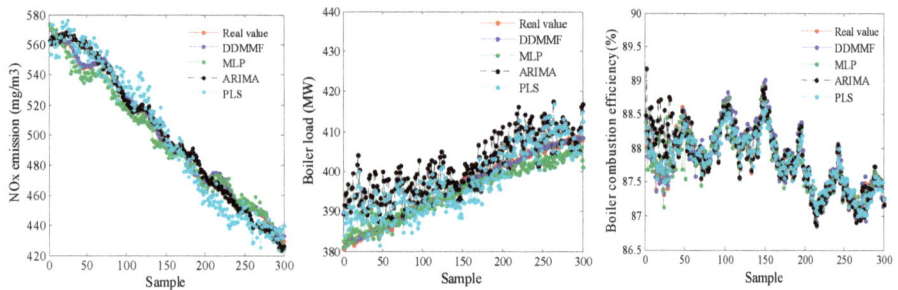

Figure 3. The predicted values obtained from different prediction models for operating condition 1.

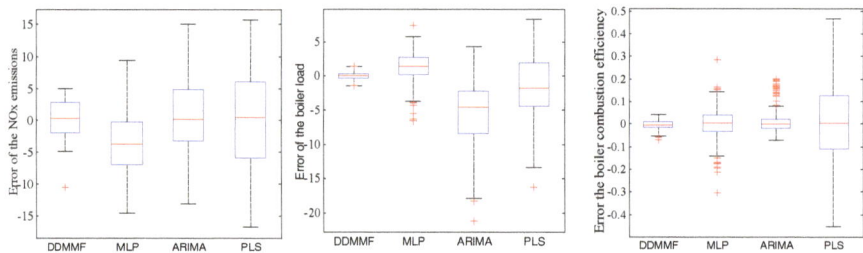

Figure 4. The prediction errors obtained from different prediction models for operating condition 1.

Table 4. The prediction evaluation indicators of different objects for different models in operating condition 1.

Model	NOx Emissions			Boiler Load			Boiler Efficiency		
	MAE (mg/m^3)	RMSE (mg/m^3)	MARE (%)	MAE (MW)	RMSE (MW)	MARE (%)	MAE (%)	RMSE (%)	MARE (%)
DDMMF	1.697	13.466	4.9×10^{-4}	0.001	4.244	0.001	0.010	0.162	7.3×10^{-5}
MLP	1.766	14.138	9.4×10^{-4}	0.029	11.567	0.011	0.058	0.184	2.0×10^{-4}
ARIMA	2.559	28.249	1.9×10^{-3}	3.520	21.563	0.789	0.040	0.712	5.1×10^{-4}
PLS	2.321	15.844	3.0×10^{-3}	0.963	35.940	0.631	0.118	1.628	2.8×10^{-4}

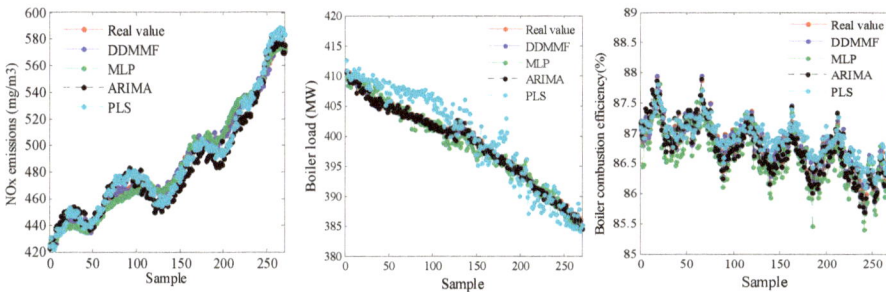

Figure 5. The predicted values obtained from different prediction models for operating condition 2.

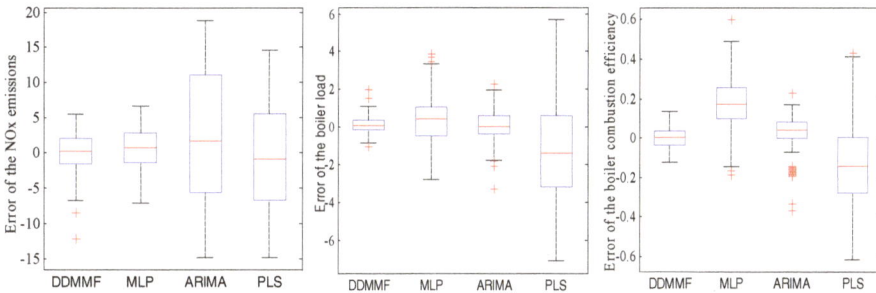

Figure 6. The prediction errors obtained from different prediction models for operating condition 2

Table 5. The prediction evaluation indicators of different objects for different models in operating condition 2.

Model	NOx Emissions			Boiler Load			Boiler Efficiency		
	MAE (mg/m^3)	RMSE (mg/m^3)	MARE (%)	MAE (MW)	RMSE (MW)	MARE (%)	MAE (%)	RMSE (%)	MARE (%)
DDMMF	1.518	1.674	0.329	0.519	1.789	0.003	0.021	0.474	1.5×10^{-4}
MLP	1.883	7.832	1.899	4.683	6.649	0.012	0.149	2.975	8.4×10^{-4}
ARIMA	5.405	15.062	1.054	0.001	6.634	0.005	0.128	0.981	3.7×10^{-4}
PLS	2.321	16.946	2.443	0.115	4.630	0.005	0.065	0.684	5.4×10^{-4}

The prediction accuracy of the DDMMF model was better than the corresponding one obtained with the MLP, ARIMA and PLS models. For example, in case of operating condition 1 compared with the other predictive models, the MAE indicator of the DDMMF model for the NOx emissions prediction reduced by 0.391%, 33.685% and 26.885%, respectively; the RMSE indicator was diminished by 4.753%, 52.331% and 15.009%, respectively; and the MARE indicator decreased by 48.320%, 74.557%

and 83.602%, respectively. The runtime of the comparison models was 46.426 s, 52.678 s and 38.124 s, respectively. And the runtime of the DDMMF model was only 27.844 s. In this case, the MAE indicator of the DDMMF model for the boiler load prediction decreased by 88.928%, 73.028% and 67.557%, respectively; the RMSE indicator reduced by 23.228%, 73.028% and 28.179%, respectively; and the MARE indicator decreased by 51.754%, 61.357% and 19.373%, respectively. The runtimes of the comparison models were 40.128 s, 43.324 s and 38.421 s, respectively. The runtime of the DDMMF model was only 30.564 s. Similarly, for the boiler combustion efficiency, the MAE of the DDMMF model reduced by 96.551%, 99.715% and 99.896%, respectively; the RMSE diminished by 63.318%, 40.707% and 35.653%, separately; and the MARE decreased by 62.630%, 94.816% and 93.517%, respectively. The runtimes of the other models were 36.345 s, 32.157 s and 38.657 s, respectively. The runtime of the DDMMF model was only 25.625 s. Similar results were obtained for operating condition 2, the MAE indicator of DDMMF model for the NOx emissions prediction diminished by 19.343%, 71.906% and 44.512%, respectively; the RMSE indicator reduced by 78.628%, 88.887% and 90.123%, respectively; and the MARE indicator decreased by 82.696%, 68.832% and 86.545%, respectively. The runtimes of other models were 36.175 s, 38.620 s and 32.265 s, respectively. The runtime of the DDMMF model was only 28.658 s. For the same operating condition and boiler load prediction, a 88.928%, 73.028% and 67.557% decrease in the MAE indicator of DDMMF model, a 23.228%, 73.028% and 28.179% decrease in the RMSE indicator and a 51.754%, 61.357% and 19.373% decrease in the MARE index was obtained, respectively. The runtimes of the comparison models were 34.659 s, 33.674 s and 35.658 s, respectively. The runtime of the DDMMF model was only 29.467 s. Finally, the boiler combustion efficiency prediction showed a 85.998%, 83.661% and 67.915% reduction in the MAE indicator of the DDMMF model, 84.067%, 51.682% and 30.702% decrements in the RMSE index and a 82.196%, 59.790% and 72.201% reduction in the MARE index, respectively. The runtimes of other models were 48.554 s, 52.214 s and 43.248 s, respectively. The runtime of the DDMMF model was only 19.248 s. The results of the present study indicate that the algorithm improved of the modeling accuracy and counting efficiency.

4. Adaptive Nonlinear Model Predictive Control of the Boiler Combustion Efficiency

The model predictive control (MPC) of the boiler combustion efficiency is described in this section. The overall control structure, including the prediction model, the rolling optimization, as well as the feedback correction is illustrated in Section 4.1. The rolling optimization and feedback correction are discussed in Sections 4.2 and 4.3, respectively. The validation of the designed model predictive controller is performed in Section 4.4.

4.1. The Structure of the Proposed Model Predictive Controller

The MPC algorithm consists of the prediction model, rolling optimization and feedback correction as shown in Figure 7. The LSSVM-based prediction model is optimized via the DE algorithm. Predictive models are provided for the NOx emissions, the boiler load, and the boiler combustion efficiency to predict the boiler combustion efficiency under the NOx emissions and the boiler load. Historical data and current production information are employed to predict the relevant object values at the current time. Then, the prediction information of the DDMMF is adopted to establish rolling optimization. The optimal controller parameters are obtained by solving this rolling optimization problem. This gives maximum boiler combustion efficiency under the NOx emissions and the boiler load constraints. Feedback correction is utilized to improve the proposed model accuracy.

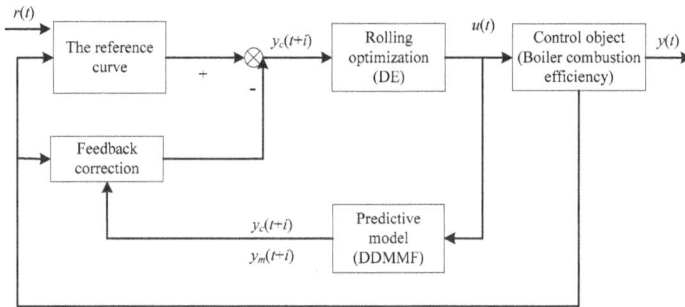

Figure 7. The proposed model predictive control structure.

4.2. Rolling Optimization

Rolling optimization is designed to obtain the optimal values of the controller parameters. The minimum control errors between the control output and expected value of boiler efficiency is taken as the control objective. The control constraints consist of three parts: (1) The error of NOx emission control output and expected output must be within 5%; (2) the error of boiler load control output and expected output must be within 5%; and (3) the control variables must satisfy the boundary constraints. The control output value of the NOx emissions and boiler load, and the optimal control variables are output when the control errors of the NOx emissions and boiler load meets the constraints and minimum requirements of the control objectives. The maximum boiler efficiency can be obtained by reasonably allocating control variables. The boiler load expected value is the actual operation value, the boiler is in a stable operation state when the boiler load control output can better follow the expected curve. As a result, the maximum boiler combustion efficiency is constructed by considering load constraints, pollutant emission constraints and boundary constraints could be obtained. An appropriate performance index is minimized to determine the control increment, such that p future output values of boiler efficiency Y_B, track the desired value of the boiler combustion efficiency ω_B. The boiler combustion efficiency is considered as the performance index of the MPC controller for achieving the rolling optimization. Thus, the following rolling optimization problem is defined:

$$
\begin{aligned}
&\min\left(\sum_{s=1}^{p}\left(Y_B - \omega_B^s\right)\right) \\
&s.t. Y_{B,h} = f(Fc, M, A_1, A_2, A_3, A_4, P_a, S, O_1, O_2, T_1, T_2, T_3, T_4, P, \sigma, w_B, C, e_i, a_i, b, B^*) \\
&\qquad Y_{Nx,h} = f(Fc, M, A_1, A_2, A_3, A_4, P_a, S, O_1, O_2, T_1, T_2, \sigma, w_{Nx}, C, e_i, a_i, b, N_x^*) \\
&\qquad Y_{L,h} = f(Fc, M, A_1, A_2, A_3, A_4, P_a, S, \sigma, w_L, C, e_i, a_i, b, L^*) \\
&\qquad Y_{Nx,h}^{M_h} \le Y_{Nx,h} \le Y_{Nx,h}^{U_h}, \; Y_{L,h}^{M_h} \le Y_{L,h} \le Y_{L,h}^{U_h}, \; u_h^{M_h} \le u_h \le u_h^{U_h}, s = 1, 2, \dots, p, h = 1, 2
\end{aligned}
\tag{12}
$$

where ω_B^s is the desired output value of the boiler efficiency under the sth minimization function. p denotes the horizon of the minimization. $Y_{B,h}$ denotes the boiler combustion efficiency predicted value in the presence of the controlincrement for the hth operating condition. $Y_{Nx,h}^{M_h}$ is the allowed minimum value limit value of the NOx emissions under the hth operating condition; $Y_{Nx,h}^{U_h}$ is the upper limit value of the NOx emissions under the hth operating condition. $Y_{L,h}^{M_h}$ is the lower bound of the load under the hth operating condition; $Y_{L,h}^{U_h}$ is the upper bound of the load under the hth operating condition. $u_h^{M_h}$ is the allowed minimum value limit value of the control variables under the hth operating condition; $u_h^{U_h}$ is the upper bound of the control variables under the hth operating condition; u is the control variable.

To solve the rolling optimization problem and find the optimal controller parameters, a DE algorithm is utilized (its flowchart is given in Figure 8). According to Figure 8, the algorithm is given as follows.

Step 1: The data is read. Then, the sampling time of the predictive model is determined to meet the rolling optimization requirements.

Step 2: The DE algorithm is initialized. The control variable is calculated as:

$$u = u_m + \sum_{i=1}^{P} d_i(\omega_B - y(i)) \tag{13}$$

Now, the particle denotes the mth control variable, including Fc, M, A_1, A_2, A_3, A_4, P_a, S.

Step 3: The N_x and the L for any particle are calculated according to Equation (5).

Step 4: If the N_x and the L requirements are satisfied, go to Step 5. Otherwise, set the relative particle fitness to 1000 and go to Step 6.

Step 5: Compute the boiler combustion efficiency of each particle. Now, calculate the fitness through the following equation:

$$\varepsilon(k) = \sqrt{(\sum_{k=1}^{p} \left|Y_B - \omega_B^k\right|^2 / (p-1))} \tag{14}$$

where Y_B denotes the predicted boiler combustion efficiency of the kth predictive mode; ω_B^k is the desired boiler combustion efficiency at time k.

Step 6: Search the minimum fitness value.

Step 7: If the stopping condition is satisfied, stop the algorithm and show the optimal controller parameters, as well as the optimal fitness value. Otherwise, go to Step 8.

Step 8: If $g = G$, show the optimal controller parameters and the fitness value and stop the algorithm. Otherwise, go to Step 9.

Step 9: Update particles according to Equations (6)–(8). Set $g = g + p$ and return to Step 3.

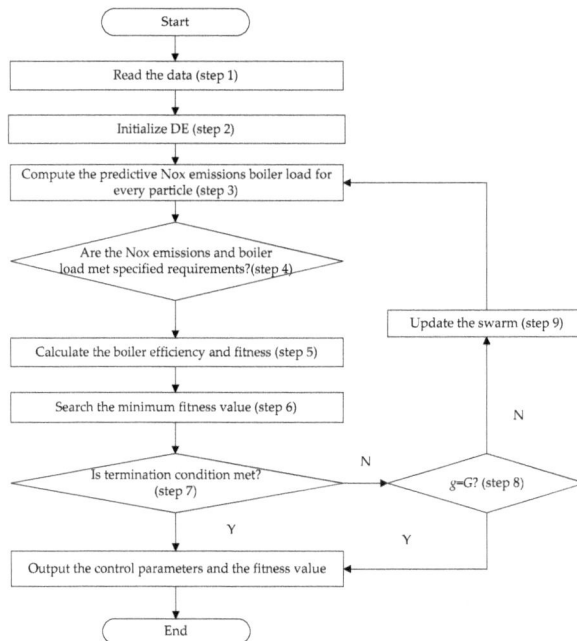

Figure 8. The rolling optimization flowchart.

4.3. Feedback Correction

The NOx emissions and the boiler load are affected by serious factors, such as fuel combustion and operating condition. Then, the model correction strategy is employed to guarantee the predictive model accuracy. In other words, recent data is applied to reconstruct the predictive model when the prediction errors of the actual NOx emissions, the actual boiler load and the boiler combustion efficiency are larger than 5%. The predictive model inaccuracy could be corrected through the model correction strategy. Therefore, the optimal results can be obtained via the model predictive control.

4.4. Validation of the Model Predictive Control Results

4.4.1. Experiment Setup

The model predictive control experimental data are shown in Table 6. The input variables of the predictive control model are obtained according to Equation (12) under two operating conditions.

Table 6. Dataset2.

Parameter	Number (Condition 1)			Number (Condition 2)		
	N_x	B	L	N_x	B	L
Test data (samples)	300	300	300	270	270	270
Sample Interval (min)	1	1	1	1	1	1
Input variable	13	16	9	13	16	9
Output variable	1	1	1	1	1	1

4.4.2. Experimental Result Analysis

In this study, the model predictive control strategy is employed to maximize boiler combustion efficiency under two operating conditions. To verify the proposed method performance, some experiments were performed using the real production data. The data sets under the two operating conditions are described in Table 7. The experimental results are given in Figures 9 and 10. The obtained results under the first and the second operating conditions are shown in Figures 9 and 10, respectively. Figure 9, Figure 10, and Table 7 demonstrate the superiority of the proposed model predictive controller (BMPC) to the proportion-integral-derivative (PID) controller. The as seen in Figures 9 and 10, with the model predictive control algorithm using in the boiler, it could be concluded that the NOx emissions control output values and boiler load control output values when the control variables were reasonably matched, which could follow the trajectory of a given curve better. In combination with Table 7, the boiler efficiency could be improved by adjusting the control values, when the NOx emissions control output errors and the boiler load control output errors were both within ±5% under the two different operating conditions. The control output error was between the predicted value under the optimal variable and the expected value. The NOx emissions and boiler load were the predicted output values of the test data using the DDMMF method. The expected values of boiler efficiency were the actual values plus a 1.5% smoothing filter, the expected values of the NOx emissions were the actual values minus 10, and the expected values of the boiler load were the actual output values. We can see from Table 7, the proposed approach compared with the PID controller could improve boiler efficiency by 1.613 percent and reduce NOx emissions by 1.887 percent for the first operating condition. In addition, it could improve boiler efficiency by 1.136 percent and reduce NOx emissions by 2.166 percent for the second condition, when the control variables satisfied the boundary constraints and the NOx emissions control output error and boiler load control output error were both within ±5%. Thus, based on the boiler key parameter model (NOx emissions, boiler load, boiler efficiency), the boiler combustion efficiency rolling optimization model was constructed by considering load constraints, pollutant emission constraints and boundary constraints, and the optimal controller parameters were calculated via an optimization model solved by the DE algorithm. Finally, the proposed BMPC provided effective tools for the operators.

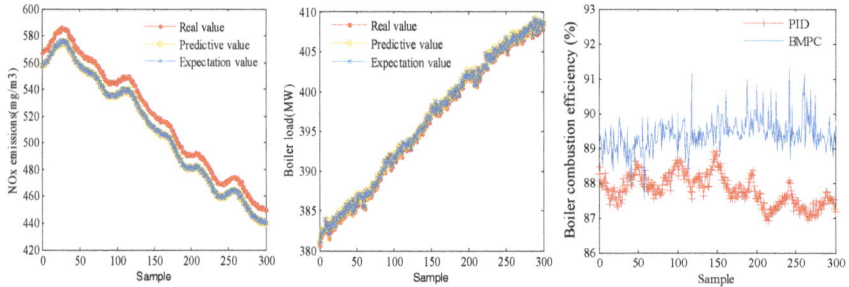

Figure 9. Theresults obtained with the model predictive control under operating condition 1.

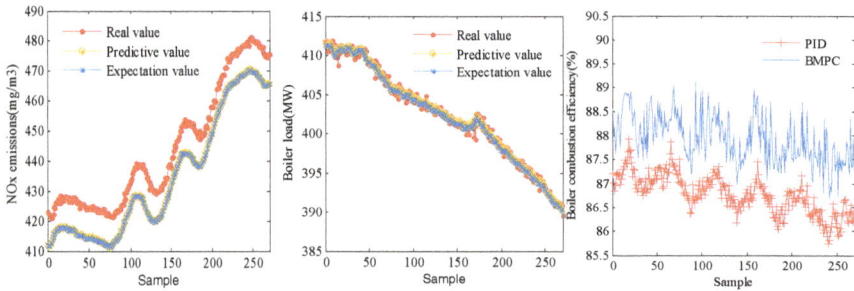

Figure 10. The results obtained with the model predictive control under operating condition 2.

Table 7. Data description.

Description	The Predictive Control Error of NOx Emissions / Boiler Load (%)	The Percentage of NOx Emissions Decline by BMPC	The Percentage of Boiler Efficiency Improvement by BMPC	Number of Samples
the operation condition 1	−4.911~4.969	1.887	1.613	300
the operation condition 2	−4.989~4.995	2.166	1.136	270

5. Conclusions

In this paper, a nonlinear model predictive control approach for different operating conditions was proposed to maximize the boiler combustion efficiency in the presence of the NOx emissions and the boiler load variations. This approach had five distinctive features: (1) PCA was employed to generate the required input for the prediction model and reduce the input dimension; (2) the KNN classifier was utilized to deal with various working conditions; (3) the LSSVM parameters were dynamically optimized by the DE algorithm, and additionally, the model output value was compensated to improve the prediction accuracy; (4) the optimal values of the control variables were obtained using the proposed adaptive nonlinear predictive control approach; (5) considering the boiler combustion efficiency, the NOx emissions, and the boiler load, the DE algorithm was employed to solve the rolling optimization problem. Experimental verification based on the practical data was carried out to verify the performance of the proposed method. According to the obtained results, both the modeling and controlling were effective. The DDMMF model could accurately predict the different objects under different operation conditions. The BMPC control strategy, based on DDMMF model, was proposed to improve the quality of production and reduce energy consumption, and to lay the foundation for a power enterprise to bring high efficiency and low emission. This provides an appropriate method that could be employed in production applications.

Future research will be dedicated to selecting the appropriate kernel function of the LSSVM under different conditions to meet the modeling requirement and improve the predictive performance of the DDMMF.

Author Contributions: Z.T. designed the research and the article structure, and revised the manuscript. X.W. carried out the experiments. S.C. revised the manuscript. All authors read and approved the final paper.

Funding: The authors would express their appreciation to the National Natural Science Foundation of China (NO.61503072 and NO.51606035) and the Science and Technology Development Plan of Jilin Province (NO. 20190201095JC, 20190201098JC).

Conflicts of Interest: The authors declare no conflict of interest.

Acronyms

NOx	Nitrogen Oxides	CFD	Computational fluid dynamics
ARIMA	Autoregressive integrated moving average	PLS	Partial least squares
ANN	Artificial neural networking	BP	Back-propagation
MLP	Multilayer perceptron	LSSVM	Least square support vector machine;
DE	Differential evolution	SIS	Supervisor information system
DCS	Distribution control system	PCA	Principal component analysis
KNN	Kth Nearest Neighbor	PSO	Particle swarm optimization
GA	Genetic algorithm	ADRC	Active disturbance rejection controller
MFC-VRFT	Model-Free control and virtual reference feedback tuning	MPC	Model predictive control
B	Boiler efficiency	L	Boiler load
N_x	NOx emissions	DELSSVM	DE-based LSSVM
MAE	Mean absolute error	MAE	Mean absolute error
RMSE	Root mean square error	MAPE	Mean absolute percentage error
DDMMF	A data-driven modeling method with feature selection capability	BMPC	Model predictive controller
PID	Proportion-integral-derivative	RBF	Radial based function

References

1. Ławryńczuk, M. Nonlinear predictive control of a boiler-turbine unit: A state-space approach with successive on-line model linearisation and quadratic optimisation. *ISA Trans.* **2017**, *67*, 475–495. [CrossRef] [PubMed]
2. Zhou, H.; Zheng, L.; Cen, K. Computational intelligence approach for NOx emissions minimizationin a coal-fired utility boiler. *Energy Convers. Manag.* **2010**, *51*, 580–586. [CrossRef]
3. Li, K.; Thompson, S.; Peng, J.X. Modeling and prediction of NOx emission in a coal-fired power generation plant. *Control Eng. Pract.* **2004**, *12*, 707–723. [CrossRef]
4. Gao, Y.; Zeng, D.; Liu, J.; Jian, Y. Optimization control of a pulverizing system on the basis of the estimation of the outlet coal powder flow of a coal mill. *Control Eng. Pract.* **2017**, *63*, 69–80. [CrossRef]
5. Ma, J.; Xu, F.; Huang, K.; Huang, R. Improvement on the linear and nonlinear auto-regressive model for predicting the NOx emission of diesel engine. *Neurocomputing* **2016**, *207*, 150–164. [CrossRef]
6. Fan, H.; Zhang, Y.; Su, Z.; Wang, B. A dynamic mathematical model of an ultra-supercritical coal fired once-through boiler-turbine unit. *Appl. Energy* **2017**, *189*, 654–666. [CrossRef]
7. Kim, B.S.; Kim, T.Y.; Park, T.C.; Yeo, Y.K. Comparative study of estimation methods of NOx emission with selection of input parameters for a coal-fired boiler. *Korean J. Chem. Eng.* **2018**, *35*, 1779–1790. [CrossRef]
8. Strušnik, D.; Golob, M.; Avsec, J. Artificial neural networking model for the prediction of high efficiency boiler steam generation and distribution. *Simul. Model. Pract. Theory* **2015**, *57*, 58–70. [CrossRef]

9. Yin, L.B.; Liu, G.C.; Zhou, J.L.; Liao, Y.F.; Ma, X.Q. A calculation method for CO2 emission in utility boilers based on BP neural network and carbon balance. *Energy Procedia* **2017**, *105*, 3173–3178. [CrossRef]

10. Liukkonen, M.; Hälikkä, E.; Hiltunen, T.; Hiltunen, Y. Dynamic soft sensors for NOx emissions in a circulating fluidized bed boiler. *Appl. Energy* **2012**, *97*, 483–490. [CrossRef]

11. Lv, Y.; Liu, J.; Yang, T.; Zeng, D. A novel least squares support vector machine ensemble model for NOx emission prediction of a coal-fired boiler. *Energy* **2013**, *55*, 319–329. [CrossRef]

12. Lv, Y.; Yang, T.; Liu, J. An adaptive least squares support vector machine model with a novel update for NOx emission prediction. *Chemom. Intell. Lab. Syst.* **2015**, *145*, 103–113. [CrossRef]

13. Xu, C.; Gao, S.; Li, M. A novel PCA-based microstructure descriptor for heterogeneous material design. *Comput. Mater. Sci.* **2017**, *130*, 39–49. [CrossRef]

14. He, F.; Zhang, L. Prediction model of end-point phosphorus content in BOF steelmaking process based on PCA and BP neural network. *J. Process Control* **2018**, *66*, 51–58. [CrossRef]

15. Sayed, M.; Gharghory, S.M.; Kamal, H.A. Gain tuning PI controllers for boiler turbine unit using a new hybrid jump PSO. *J. Electr. Syst. Inf. Technol.* **2015**, *2*, 99–110. [CrossRef]

16. Anarghya, A.; Rao, N.; Nayak, N.; Tirpude, A.R.; Harshith, D.N.; Samarth, B.R. Optimized ANN-GA and experimental analysis of the performance and combustion characteristics of HCCI engine. *Appl. Eng.* **2018**, *132*, 841–868. [CrossRef]

17. Han, J.; Wang, H.F.; Jiao, G.T.; Cui, L.M.; Wang, Y.R. Research on active disturbance rejection control technology of electromechanical actuators. *Electronics* **2018**, *7*, 174. [CrossRef]

18. Roman, R.C.; Radac, M.B.; Precup, R.E.; Pertriu, E.M. Virtual reference feedback tuning of model-Free control algorithms for servo systems. *Machinces* **2018**, *5*, 25. [CrossRef]

19. Pan, Q.K.; Suganthan, P.N.; Wang, L.; Gao, L.; Mallipeddi, R. A differential evolution algorithm with self-adapting strategy and control parameters. *Comput. Oper. Res.* **2011**, *38*, 394–408. [CrossRef]

20. Yu, X.; Yu, X.; Lu, Y.; Yen, G.G.; Cai, M. Differential evolution mutation operators for constrained multi-objective optimization. *Appl. Soft Comput.* **2018**, *67*, 452–466. [CrossRef]

21. Afram, A.; Janabi-Sharifi, F. Supervisory model predictive controller (MPC) for residential HVAC systems: Implementation and experimentation on archetype sustainable house in Toronto. *Energy Build.* **2017**, *154*, 268–282. [CrossRef]

22. Balram, P.; Tuan, L.A.; Carlson, O. Comparative study of MPC based coordinated voltage control in LV distribution systems with photovoltaics and battery storage. *Electr. Power Energy Syst.* **2018**, *95*, 227–238. [CrossRef]

Article

Integrated Approach for Network Observability and State Estimation in Active Distribution Grid

Basanta Raj Pokhrel *, Birgitte Bak-Jensen and Jayakrishnan R. Pillai

Department of Energy Technology, Aalborg University, 9220 Aalborg, Denmark; bbj@et.aau.dk (B.B.-J.);
jrp@et.aau.dk (J.R.P.)
* Correspondence: bap@et.aau.dk

Received: 14 May 2019; Accepted: 10 June 2019; Published: 12 June 2019

Abstract: This paper presents a unique integrated approach to meter placement and state estimation to ensure the network observability of active distribution systems. It includes observability checking, minimum measurement utilization, network state estimation, and trade-off evaluation between the number of real measurements used and the accuracy of the estimated state. In network parameter estimation, observability assessment is a preliminary task. It is handled by data analysis and filtering followed by calculation of the triangular factors of the singular, symmetric gain matrix using an algebraic method. Usually, to cover the deficiency of essential real measurements in distribution systems, huge numbers of virtual measurements are used. These pseudo measurements are calculated values, which are based on the network parameters, real measurements, and forecasted load/generation. Due to the application of a huge number of pseudo-measurements, large margins of error exists in the calculation phase. Therefore, there is still a high possibility of having large errors in estimated states, even though the network is classified as being observable. Hence, an integrated approach supported by forecasting is introduced in this work to overcome this critical issue. Finally, estimation of the trade-off in accuracy with respect to the number of real measurements used has been evaluated in order to justify the method's practical application. The proposed method is applied to a Danish network, and the results are discussed.

Keywords: grid observability; active distribution system; meter allocation; parameter estimation

---

1. Introduction

Power flow in distribution networks has traditionally been one-way towards the consumers from the substation. Now the scenario is changing, and networks are becoming bidirectional. This is due to substantial demand responses programs and interconnection of time-varying distributed generation (DG) at the distribution network (DN) level (medium voltage (MV) and low voltage (LV)) [1]. Such active power system networks can be controlled properly only when the actual system state is known. In addition, there is a threat to network operators from bidirectional power flow causing operational issues for network security and voltage balance [2], which forces network operators to carry out security assessment. For this objective, network states (voltage magnitudes (V) and angles (θ) at each bus), should be evaluated under all operating conditions [3]. Based on the predicted network states, the other network parameters can be calculated, and the operators can run control modules and handle further operation effectively. To ensure the availability of estimated states, power industries today use observability analysis. However, the accuracy of the estimated states cannot be guaranteed due to a lack of sufficient real measurements and error in load/generation forecasting, especially in distribution grids. Topology change and flaws in data communication are some of the culprits for measurement inadequacy. The consequence of this is network un-observability, but the observability of this unobserved network can be reinstated by using some defined measurements [4].

Any network can be classified as observable if it is possible to calculate all system states for the given measurements and topology. There are two predominant approaches for observability analysis: (a) topological: based on the graph theory analysis and decoupled measurement model and (b) numerical: based on the numerical factorization of gain matrices using either coupled or decoupled measurement models [5]. In addition to these two methods, some other alternative approaches have also been applied for network observability assessment in power distribution grids. For instance, by defining the probability index, local observability assessment using graph theory criteria etc. [6]. The method of distribution grid observability enhancement using a smart meter data is explained in [7]. This method uses both time-varying injections and stationary conventional load data from the smart meter as metered and non-metered data for observability improvement. State estimation in this case is executed using the metered data from a small number of buses to solve the non-linear power flow equations over consecutive time instants. Observability of the network depends on number, type and locations of the measurements used, but not on the method followed. Generally, there are sufficient measurements available in the transmission network that these methods work well there. However, due to economic reasons, there are fewer measuring devices (e.g., power, current and voltage measurements) in distribution systems, especially at the MV level, so they are generally operated with reduced observability. On the other hand, in the emerging scenario, the use of smart meters at the LV level (every customer connection point) is rapidly increasing because of the availability of modern advanced metering infrastructures (AMI) [1]. Some of the examples of increasing measurements at the distribution level are: (a) the European Union's (EU) initiative to replace 80% of traditional electricity meters with smart meters by 2020 [4], (b) installation of more than 70 million (nearly half of U.S. electricity customers) AMI in the United States (U.S.) [8], etc. Therefore, in future there will be a huge amount of data available from almost every node in the LV distribution grid, and they will be stored in data centers such as data hubs. These data can be used for any applications. For control applications, these large amounts of data gathered by smart meters have to be processed and used in an economical way. Hence, smart selection of minimum key data from the huge pool of available data from customer locations as well as from other specific locations will enhance the observability of active distribution networks (ADNs), while at the same time, ensuring higher accuracy of the estimated states of the observable network is the main challenge [7]. The Fisher information-based meter placement technique by minimizing errors in estimated state vector is proposed in [9]. It uses the D-optimality criterion, and a Boolean-convex model is used for optimization of the problem formulation. A meter placement algorithm to improve the estimated voltage and angle at each bus in the network is described in [10]. This technique is based on the consecutive upgrading of a bivariate probability index that is applied in medium voltage networks. The minimum meter placement technique discussed in this paper is a simple algebraic technique that can be applied in both MV and LV control applications. For MV grids, it will determine the best minimum metering locations, and for LV grids, it will identify the minimum key data from the huge pool of smart meter data. Using these smartly selected data, the respective networks can be fully observed with an accurately estimated network state from all nodes.

A reduced observability scenario below the substation is one of the reasons for the limited use of distribution state estimation (DSSE) by the utility. However, now, to address the challenges resulting from the increasing penetration of flexible resources and DGs at DN, real-time network models based on DSSE are becoming more essential for operation and control of the DN [11]. In the available literature, many vital issues in the development of DSSE have been discussed [12–14]. Proposed algorithms have been categorized based on simplification in order to speed up the estimation, selection of the state variables, and the procedures for integrating diverse measurements. Considering the choice of state variables, branch current-based and node voltage-based state estimators are the two main categories that would ultimately results in similar accuracy at the end [15]. The WLS (weighted least square) method's performance analysis with respect to the choice of state variables and comparisons on the basis of accuracy, performance and bad data detection capability were reported in [16]. To use the estimated network state information for operation and control, state estimation has to be dynamic, irrespective of the method used for estimation. Dynamic state estimation is a recurrent estimation in a time sequence based on several measurement snapshots. For the estimation of large distribution networks, the execution time will be longer if we assume a single estimation process for the whole network. Therefore, distributed DSSE methods can be used to overcome such issues, executing the DSSE module in different sub-areas of the distribution network locally. Network division for this purpose can be carried out according to topological framework, geographical standpoint or measurement locations to solve the problem via local estimators in that particular sub-area. However, it is challenging to use DSSE in real life networks due to the limited number of real measurements, delay in communication, and un-synchronized measurements, which are detrimental to multi-area estimation accuracy [15]. The Distribution System State Estimation procedure with multi area architecture is discussed in [17]. It is based on a two-step procedure, i.e., local estimation followed by global estimation by using integrated measurement information from adjacent areas. The WLS estimation principle is used to identify the impact on the estimation accuracy of sharing the measurements among different areas. To monitor the dynamic behavior of the power system, for example, in order to know the quasi real-time operating condition in case of any fast-evolving contingencies, phasor measurement unit (PMU) measurements are included in the measurement set used by the state estimation [18]. PMU measurements comprise a voltage phasor at a bus and a current phasor through the line incident to that bus, and have sampling synchronized to a common reference through a GPS signal at widely spread locations [19]. A number of approaches can be found in the literature to include PMU data set into the existing conventional measurement setup [20]. If only voltage and current measurements from PMU are used, the state estimation problem will be linear, but it will be nonlinear with the presence of both PMU and conventional measurements. Therefore, to linearize the problem, the measured phasors in polar form can be converted to an equivalent rectangular form. Key challenges due to the inclusion of PMU measurements include: false data injection attack in the wide area measurement system [21], convergence problem of the hybrid state estimator during the iterative process due to uncertainty introduced by the instrument transformers, their connection with digital equipment and A/D converters [22], etc. On the other hand, to overcome the limitation of real measurements, when large number of virtual measurements (that may have significant uncertainties) are used to make the network observable, there is a high possibility of system states deviating from their actual values. Hence, the optimal number and specific type of real measurements collected from specific locations are essential to address the necessities of applications for real-time operation with high accuracy [23,24]. Therefore, the challenges to develop new real-time monitoring and management solutions for smart grid still exists, such as:

1. Accurately observing the whole distribution network using minimum real data and maximum pseudo measurements.
2. Developing an integrated approach for accurate, adaptive and efficient DSSE methodologies equipped with minimum measurement technique for wide area monitoring that can be implemented in the active distribution network.
3. Identifying the practical trade-off between network observability, number of installed measuring devices, and accuracy of estimated states.

These challenges related to measurement data utilization and its trade-off in DMS to enhance the state estimation are addressed in this paper. The main contributions of the paper are: development of a unified network observability assessment technique using minimum measured data, development of a technique to identify critical measuring locations for SE based on a bus prioritization approach, and formulation of a procedure to evaluate the trade-off between SE accuracy and the number of real measurement devices to be considered. This enables the user to maintain estimation precision in the distribution network with a lower burden. Overall, the paper is structured as follows: the problem statement is described in Section 2 and the overall methodology and algorithm is explained in Section 3, which covers observability assessment, meter placement analysis, network parameter estimation and accuracy trade-off analysis between the number of used measurements and the accuracy of estimated states. Improved forecasting for pseudo measurements of loads/generations are stated in Section 4. Case studies with results and discussion are presented in Section 5. The conclusions of the paper and the directions of future work are finally summarized in Section 6.

2. Problem Statement

Network topology, grid parameters, field measurements and system data correlation are never fixed; everything depends on the operating conditions. Due to these uncertainties, network modelling is never perfect, and inaccuracies are contained in the DSSE results. When the database is extremely large, network reduction may help to improve the estimation accuracy to some extent. Key essential tasks for the enhancement of observability and improvement of accuracy of online models are: synergize the massive diverse data in different data formats from several information systems, synchronize the polling cycles, and minimize the communication delay and variation in measurements [25]. Generally, three types of measurements are considered in DSSE [15].

1. Equipment connectivity status received from Geographic Information System (GIS).
2. Real-time measurements from distribution SCADA systems (e.g., voltage, current and power flow) embedded with measuring units (e.g., smart meters, PMU, remote terminal unit (RTU), AMI, etc.).
3. Customer/prosumer's demand and DER output data (active and reactive power) from data management system.

If all of these numerous measurements are used, it may give better accuracy, but will require additional bandwidth for a communication infrastructure with suitable reliability. This will then lead to other problems, like data overload and extra financial burden. On the other hand, limited use of real-time measurement cannot ensure the overall observability of the network [26,27]. To minimize the use of real measurements, a large number of pseudo and virtual measurements can be used. Load and generation forecasts are pseudo measurements, whereas zero voltage drops in closed switching devices, zero bus injections at the switching station nodes, etc., are virtual measurements. However, changes in the behavior of power consumption because of external factors like new tariff structures, environmental condition, etc., and the allocation of greater weight to virtual measurements and lower weight to pseudo measurements may lead to ill-conditioned systems [15]. Therefore, this problem is tackled in this paper by utilizing maximum pseudo measurements only (injection measurements calculated from forecasted load/generation profiles) with a trade-off analysis where DSOs can achieve observability in their distribution grids with higher accuracy in estimated states by using minimal real measurements. The overall workflow for this problem is presented in Figure 1. As shown in Figure 1, for the particular network parameter to be estimated (in this work voltage in each bus), the type and combination of measurements (PV or PQ or IV or PQI, etc.) required are identified based on a preliminary assessment based on the available network topology and information. The best combination of measurements for real networks is identified in this preliminary assessment using the method explained in [28], and consists of analyzing the impact of number and placement of smart meter data on errors in state estimation using a standard network. This is a onetime exercise, and there is no need for repeated analysis of the preceding calculations. This is the first step, and this will identify the key measurement parameters and their locations, to a certain extent. It is followed by the observability assessment of the network with minimum meter placement, which is described in Section 3, below. Once the network is identified as being observable with minimum measurements, the state estimation (SE) is triggered. For the SE, real data from the identified key meters, together with the pseudo measurements for the rest of the nodes (which have no real measurements available), are supplied. Finally, a trade-off analysis between the number of real measurements used, the level of network observability and state estimation accuracy is carried out. For the estimation of network parameters, the weighted least square-based approach based on the Newton-Raphson method is applied. A detailed explanation of the methodology followed for each working block represented in Figure 1 is explained in Sections 3 and 4 below.

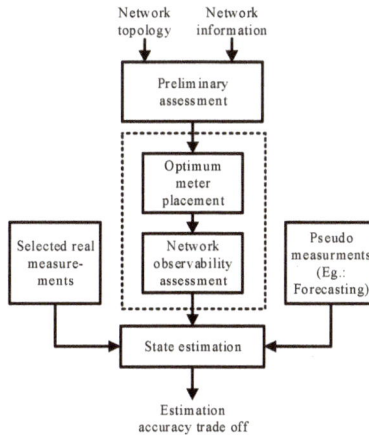

Figure 1. Working flow diagram of the proposed model.

3. Observability Assessment

The first stage in the assessment of observability is to choose estimation variables, and the best combination of parameters to be measured. This is termed preliminary assessment in this paper. It will minimize the iteration cycle in SE and give the minimum estimation error to a certain extent. For voltage estimation, line flow measurements are found to be more crucial [28]. It depends on network type and the variables to be estimated. As described in [28], if the parameters to be measured from specific locations are selected properly, better estimation quality can be achieved even with fewer real measurements. As explained above, once the estimation variables and the proper combination of measurements of parameters are selected, the actual network observability assessment will start, followed by state estimation. If the network is identified as being unobservable, all branches in it that cannot be observed have to be determined first. Then, by removing unobservable branches, the network can be divided into many individual observable islands that can be treated individually and merged later to have a single observable island. To improve the overall observability of the whole network, further analysis on the determination of specific measurement locations and the selection of minimum measurements has to be carried out [29]. The proposed integrated algorithm for observability assessment and its application in SE is given in Section 3.

Algorithm for Network Observability, SE and Accuracy Trade-Off

Algorithm for Observability Check (AOC):

- **Step 1:** Calculate measurement Jacobian matrix (H) and its gain matrix (G) [5,29].
- **Step 2:** Perform triangular factorization of G and evaluate its triangular factors (L: lower factor). Using triangular factors, identify diagonal matrix (D) [5].
- **Step 3:** Test for zero pivot in D, and if D is equal to one, network is observable [29]. If network is observable proceed to state estimation for this go to 'step one in Algorithm for SE formulation'; otherwise go to step four.
- **Step 4:** Determine test matrix W using equation: $L^T W^T = e_i$ and calculate another test matrix C using equation: $C = AW^T$ where A is incidence matrix of branches and buses.
- **Step 5:** Identify unobservable branches corresponding to the row in 'C' with at least one non-zero element. Remove all unobservable branches

- **Step 6:** Prepare the list for probable measurement points. Prioritize them. Highest priority for placing a meter is assigned to the bus that has the maximum number of incident lines in the unobservable region. Line flow measurement from the branches between the observable islands and the injection measurements from the buses at the boundary of these islands are the next most suitable candidate measurements, which can merge the islands. Current flow measurement at the beginning of the feeders are another prioritized measurement, and feeders with greater length and higher number of branches are of higher priority among the others. Select the measurement points based on priority, and go to step one for all new measurements added.

Normally in MV distribution systems, there is a limitation due to economic concerns on the availability of adequate real measurements from each node; therefore, a large number of pseudo measurements are used for the analysis. Therefore, even though the network is identified as being observable, there is a high chance that an estimated state can differ significantly from the actual state. Hence, re-examination of network observability considering the accuracy of the estimated states is recommended.

State Estimation Formulation:

SE is the core of the security analysis function in power systems. It acts like a filter between raw measurements and application functions like control and protection modules. Based on the available measurement sets, it estimates the network status. The state of the network is defined by voltage magnitude (V) and angle (θ) at every bus (i.e., 2n state variables) for an 'n'-bus power system network [5,30]. The basic algorithmic steps for network state estimation are given below:

SE Algorithm [5,30]:

- **Step 1:** Represent the network model by state vector as shown in Equation (1). If bus 1 is considered to be the reference, it is set as ($\theta_1 = 0$).

$$x^T = [\theta_2, \ldots, \theta_n, V_1, \ldots, V_n] \tag{1}$$

The measurement vector (z) is related to the state vector (x) by a nonlinear function (h) and a vector of measurement error (e) as given in (2). These functions and their behavior are dependent on network topology and actual power flow.

$$z = h(x) + e \tag{2}$$

- **Step 2:** Define the objective function as represented in Equation (3) and minimize it for the network estimation. To minimize this objective function, WLS—the most commonly used method—is used in this paper. Here, R, σ^2 and m represent the measurement error covariance matrix, measurement variance and the number of measurements, respectively.

$$\min J(x) = [z - h(x)]^T \cdot R^{-1} \cdot [z - h(x)] \tag{3}$$

$$R = \text{diag}[\sigma_1^2, \sigma_2^2, \ldots \ldots, \sigma_m^2] \tag{4}$$

This approach determines the variance of the estimated state variables by:

$$\text{cov}(x) = [H(x)^T \cdot R^{-1} \cdot H(x)]^{-1} \tag{5}$$

$$H = \left[\frac{\delta h(x)}{\delta x}\right] \tag{6}$$

Where the diagonal elements of cov(x) represent the variance of the estimated state variables.

- **Step 3**: Once the all the network states are estimated, accuracy is evaluated based on the number of measurements used. Go to 'step one of algorithm for accuracy trade-off in state estimation'.

Algorithm for accuracy trade-off in state estimation (AAT):

- **Step 1**: Choose the desired confidence level (CL). This represents the risk that the true value goes beyond the boundary of the confidence interval (CI), i.e., the lower the CL, the higher the risk, and vice versa [31].
- **Step 2**: Calculate the probability density function (PDF) of the estimated states using Equation (7) [5].

$$F_i(a_i) = \frac{1}{\sqrt{2\pi\sigma_i^2}} e^{-\frac{(a_i - E_i)^2}{2\sigma_i^2}} \tag{7}$$

Where I is the total number of network parameters that has to be estimated, and for each network parameter '*i*', its estimated value is E_i for i = [1, 2, ..., i, ..., I]. σ_i^2 represents the variance of '*i*', whereas a_i denotes a possible value of this parameter '*i*'. There are many methods for calculating the PDF of an estimated parameter in order to calculate CI. Gain matrix-based approaches assuming Gaussian distribution have been used in this work, since the formulated observability assessment method is independent of the PDF calculation method [31,32].

- **Step 3**: Calculate confidence interval (CI) for each defined PDF. For predefined values of CL, the CI end points, a_i min and a_i max, of the estimated parameter i have to satisfy Equations (8) and (9) [31].

$$\int_{a_i\,min}^{E_i} F_i(a_i).da_i = \frac{CL}{2} \tag{8}$$

$$\int_{E_i}^{a_i\,max} F_i(a_i).da_i = \frac{CL}{2} \tag{9}$$

The CI can be calculated by multiplying the standard deviation by the coverage factor related to the predefined CL. a_i min and a_i max are the endpoints of the CI and represent the accuracy of the estimated value of parameter i. Then, calculate the maximum expected difference between E_i and its true value as given by Equation (10).

$$\text{Maximum expected difference between } E_i \text{ and its true value } = |a_i max - a_i min| \tag{10}$$

- **Step 4**: Access the estimation accuracy. If Equation (10) is satisfied, the estimation can be classified as accurate and stop the process. Otherwise, go to step five.
- **Step5**: Redefine the measurement placement with next possible measuring option, change the ratio of real and pseudo measurements, and go to 'step six of algorithm for observability check'.

The overall methodology discussed in Section 3 in the form of different algorithms can be combined to form a compact flowchart as shown in Figure 2, systematically interlinking all the mathematical models (Equations (1)–(10)) for the complete analysis.

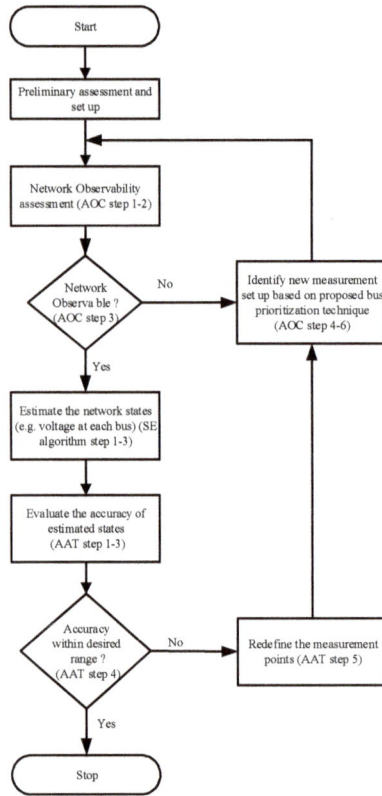

Figure 2. Compact algorithm for integrated assessment.

4. Improved Forecasting for Pseudo Measurement Modelling

Pseudo measurements are used in the network parameter estimation process to minimize the use of real measurements. For instance, even if real measurements are not sufficient to observe the network, its observability can be established by utilizing the calculated pseudo measurements. However, the accuracy of estimated states in this case depends on the accuracy of the pseudo measurement models. Pseudo-measurements here refer to forecasted load/generation values based on previous real measurements and other parameters like weather, type of load, etc. An upgraded short-term forecasting technique is proposed in this work for higher accuracy to determine the active demand and flexible generation capacity. This method considers weather forecast, historic load/generation statistics, and social and technical events. Due to the capability for handling both linear and non-linear relationships, and the facility for learning these relations directly from the data being modelled, the artificial neural network (ANN) technique is used in this work [33,34]. To obtain more accurate load forecasts, one may have to consider some other factors other than the load history that can influence the use of load by the consumers, such as time factors, weather data, possible customer classes, etc. A correlation (R) analysis is carried out to identify the most influential factor in the load forecasting using Equation (11). Here, p and q are variables to be correlated and $\bar{p}, \bar{q}$ are their respective means. For details on forecasting models and analysis, please refer to [33].

$$\text{Correlation (R)} = \frac{\Sigma(p - \bar{p})(q - q)}{\sqrt{\Sigma(q - \bar{q})^2 (p - \bar{p})^2}} \tag{11}$$

5. Case Study

The methodology described in Sections 3 and 4 is simulated for a real network and is discussed in the following sections. To demonstrate the applicability of the proposed methods, case studies have been performed on a model of a 52-bus MV distribution network from the Lind area in Denmark. For simplicity, the actual network is reduced to a 30-bus network preserving its radial nature and feeder connections, as given in Figure 3. While reducing the network, 24-load points are lumped up, i.e., five loads to bus3, three loads to bus 11, one load to bus 12, three loads to bus 13, two loads to bus 14, one load to bus 17, four loads to bus 18, and five loads to bus 19. Therefore, with this setup, in the test network, the maximum load of 1289 kW is at bus 19 and next largest load size (1246 kW) is at bus 18, while the minimum load of 18 kW is at bus 16. The typical R/X ratio of distribution lines in this network varies between 7.07 to 1.38.

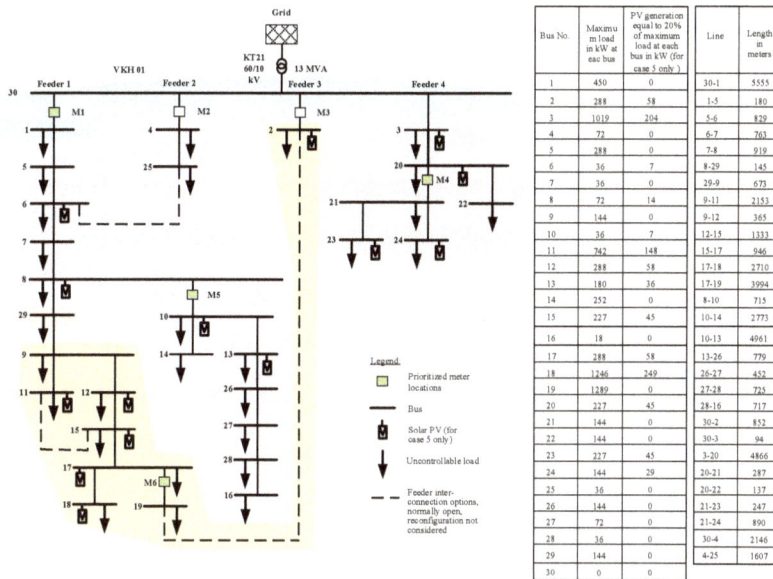

Bus No.	Maximum load in kW at each bus	PV generation equal to 20% of maximum load at each bus in kW (for case 5 only)	Line	Length in meters
1	430	0	30-1	5555
2	288	58	1-5	180
3	1019	204	5-6	829
4	72	0	6-7	763
5	288	0	7-8	919
6	36	7	8-29	145
7	36	0	29-9	673
8	72	14	9-11	2153
9	144	0	9-12	365
10	36	7	12-15	1333
11	742	148	15-17	946
12	288	58	17-18	2710
13	180	36	17-19	3994
14	252	0	8-10	715
15	227	45	10-14	2773
16	18	0	10-13	4961
17	288	58	13-26	779
18	1246	249	26-27	452
19	1289	0	27-28	725
20	227	45	28-16	717
21	144	0	30-2	852
22	144	0	30-3	94
23	227	45	3-20	4866
24	144	29	20-21	287
25	36	0	20-22	137
26	144	0	21-23	247
27	72	0	21-24	890
28	36	0	30-4	2146
29	144	0	4-25	1607
30	0	0		

Figure 3. Modified MV network of Lind area, Denmark (observable islands in case 3).

The types of load and the state of the transformer tap changer will have an impact on network performance. For simplicity, an operating scenario in which all distribution transformers (buses 1 to 29) are loaded with a capacity factor of 40% and load power factor of 0.9 is assumed for the load flow and network measurements. The load distributions in all of the buses in this scenario are given in the table included in Figure 3. This operating condition is the peak load scenario for the main transformer (KT21). Another assumption is the presence of solar PV, which is considered to be available only in case 5.

Initially (case 1), it is assumed that 14 measurements from the substation (bus 30), the far end of the feeders (buses: 2, 11, 14, 16, 18, 19, 22, 23, 24, 25), and lines (6–7, and 17–19, i.e., close to the substation and at the far end of the longest feeder) are available. Thus, measurements are available for eleven power injections (net active/reactive power at the bus that can be obtained via load generation forecasting), two line flows (active/reactive power), and one voltage (at the substation). As per this measurement configuration, the network observability is analyzed by determining the gain matrix and its triangular factors. The upper triangular factor of the measurement gain matrix is evaluated and its diagonal elements (D) are identified. It is found that 28 elements out of 30 in D are zero. This shows the network is unobservable, and it is then divided into ten observable islands by removing the unobservable branches, as shown in Table 1. The candidate measurements list is prepared in consideration of the boundary of observable islands. To justify the requirement of bus prioritization, initially, power injection measurements close to buses 3, 6, 12, 4, 27, and 29 are selected arbitrarily from the list as case 2. After each added measurement, observability assessment is carried out. The upper triangular matrix is updated and its new diagonal elements (D′) are found, in which 27 zero pivots are observed. Since significant improvement has not been achieved, the network is divided into eight observable islands. Now, the candidate measurements are prioritized according to the description given in step 6 of the observability algorithm in Section 3. Only 10 measurement locations (5, 7, 8, 10, 15, 17, 20, 21, 25 and 30) out of 16 candidate sites are considered, and those remaining are filtered out. In reality, both power/current flow and power injection measurements can be measured from the same smart meter. Based on the user's requirements, the smart meters can be tuned to measure specific parameters. This option is available in most of the smart meters available today. Therefore, for the new injection measurements to be added for case 3, locations 7 and 15 from the priority list, which are close to the branch flow measurements, are selected. Also, line 30–1, and 20–21 for the branch flow measurement are considered to be added as per priority for case 3. After these measurements have been added, the new upper triangular matrix is evaluated, and updated diagonal elements (D″) are found. It is noticeable that observability is improved in this case due to significantly reduced number of zero pivots. However, the network is not fully observable. In this case, five observable islands are identified. Next, in case 4, the candidate list is updated based on the observable islands in this case, and prioritized as 5, 8, 10, 15, 20, 25, 29 and 30. Injection at locations 15 and 20 from the boundary of the islands that are observable are now selected, together with the branch flow in lines 30–1, 20–21, 8–10 and 17–19 (M1, M4, M5 and M6), as per priority. The diagonal elements of the upper triangular matrix (D‴) from the updated triangular factors are calculated. In matrix D‴, only one zero pivot can now be observed. This shows that the network is now fully observable. These case studies for network observability are summarized in Table 1. Once the network observability has been assessed, network states are estimated for each case using selected real measurements and pseudo measurements (the forecasted load and generation value at each bus that will give net injection measurements at the respective buses). This is followed by accuracy trade-off evaluation. The case studies for state estimation and accuracy trade-off evaluation carried out after the network observability assessment are presented in the subsequent Sections 5.1 and 5.2.

Table 1. Observability assessment results.

Cases Studies	% of Real Measurements Applied	Observable Islands	Remarks
Case 1 (C1): Initial Case	0	{30}, {1}, {5 6 7 8 25 4 10 14 13 26}, {27 28}, {16}, {9 11 12 15 17 18 19}, {2}, {3}, {20 21 22 23 24}, {29}.	Network Unobservable
Case 2(C2): Adding selected measurements without priority	30	{30}, {1}, {5 6 7 8 25 4 10 14 13 26 27 28}, {16}, {9 11 12 15 17 18 19}, {2}, {3 20 21 22 23 24}, {29}.	No significant improvement
Case 3(C3): Adding selected measurements as per priority	22	{30 1}, {5 6 7 8 25 4 10 14 13 26 27 28 16}, {9 11 12 15 17 18 19 2}, {3 20 21 22 23 24}, {29}.	Significant improvement
Case 4(C4): Adding selected measurements as per priority	26	{1 2 3 4 5 6 7 8 9 10 11 12 13 14 15 16 17 18 19 20 21 22 23 24 25 26 27 28 29 30}	Fully observable network

5.1. Parameter Estimation and Accuracy Trade Off

Network observability setups with the minimum number of identified measurements can only be useful if this configuration can be used to estimate network states close to the actual values. Therefore, for instance, voltage magnitude is estimated for each case and compared with actual measured values, as shown in Figure 4 for selected buses close to the substation (buses 1 and 2), the far end (buses 19 and

24), and the middle (buses 13, 15, 20 and 25) of the feeders. In the measurement a minimum voltage of 0.88 pu is observed at bus 18 when all feeder interconnection options (6–25, 11–15 and 2–19) are open, i.e., at the end of the first radial (the longest radial feeder in the setup), which violates the grid code [35]. Buses 13 and 15 are also in the same feeder as bus 18, but are towards the substation, so they have slightly more voltage than in bus 18. Since buses 25, 2, 20, and 24 are in the 2nd, 3rd and 4th feeders (short feeders, i.e., lightly loaded compared to the first feeder), respectively, they have higher bus voltages (close to 0.99 pu). In real network operation, reconfiguration is used to maintain the grid code (e.g., minimum voltage of 0.95 at bus 16 while closing 2–19 and opening 9–12), but reconfiguration is not considered in this paper. In addition, a fixed load of 40% is considered in this work to see the performance of the proposed algorithm in special cases (case 1–7). This loading scenario may not exist all the time in real network operation. The results are obtained from a series of repeated simulations in MATLAB (observability analysis) and DigSILENT power factory (load flow, measurement setup, and state estimation) and snapshots of the results are shown here. The accuracy class of all smart meters is considered to be 0.5, as per IEC 62053-11 [36]. In this simulation, bus 30, which is close to the main substation, is considered to be the reference bus, the measured voltage of which is 0.99 pu. The first feeder is extremely long compared to the other three feeders, so lower voltages are observed in the buses (e.g., in buses 1, 13 15 and 18) in this feeder. Meanwhile, the other three feeders are short compared to the first feeder, so the voltages in the buses (e.g., in buses 2, 20, 24, 25) in these feeders are almost the same as the reference voltage, i.e., 0.99.

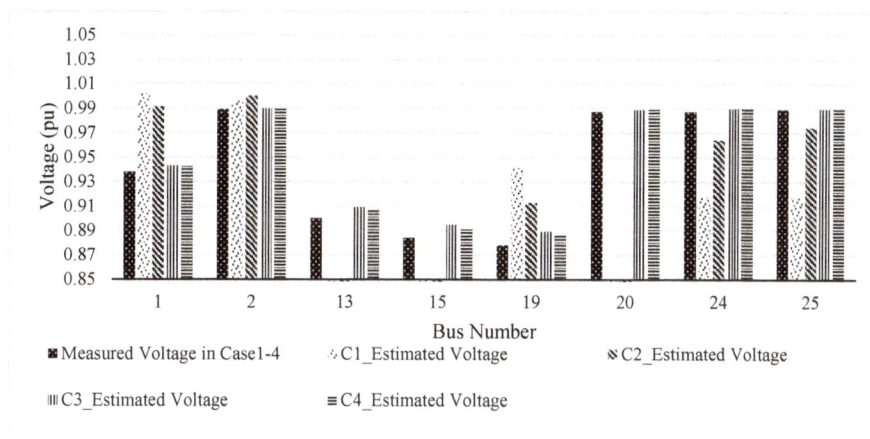

Figure 4. Comparison of estimated voltages for cases 1–4.

As expected, in case 1, where measurements are insufficient, the network is not fully observable, i.e., it is not possible to estimate the bus voltage for all buses (e.g., buses 13, 15 and 20), as is shown in Figure 4, where the magnitude of the estimated voltage for these buses is not available. In this case, overall network observability, which is a prerequisite for SE, is established by reducing the number of control variables (suspending the states to be estimated for buses 13, 15 and 20), and the bus voltages for rest of the buses is estimated. In case 2, real measurements (30%) are supplied, but without any bus prioritization, resulting in there being no significant improvement in the network state estimation. However, after applying the bus prioritization technique, i.e., by selecting only key measurements from the selected buses (only 22% real measurements) as in case 3, significant improvement is achieved. All network states including bus 13 and 15 are estimated due to the key added measurements from the prioritized buses. In case 4, which has 4% more real measurement than in case 3, all network states are estimated, with estimated states being closer to the real values. This verifies the network observability assessment results presented in Table 1. Comparison of parameter estimation errors in different case studies for the selected bus voltages are shown in Figure 5. As seen here, it is a trade-off between higher accuracy and investment in measuring devices. The magnitude of error shows the deviation of estimated quantity from its respective measured value. Positive error means overestimation and negative error means underestimation. The magnitude of error depends upon the number, type and location of the measurements considered for the respective case studies. In case 3, a maximum error of 1.24% (in bus 19) using fewer real measurements can be achieved. Also, further improvement is observed with a maximum error of 0.96% in the estimated voltages in case 4.

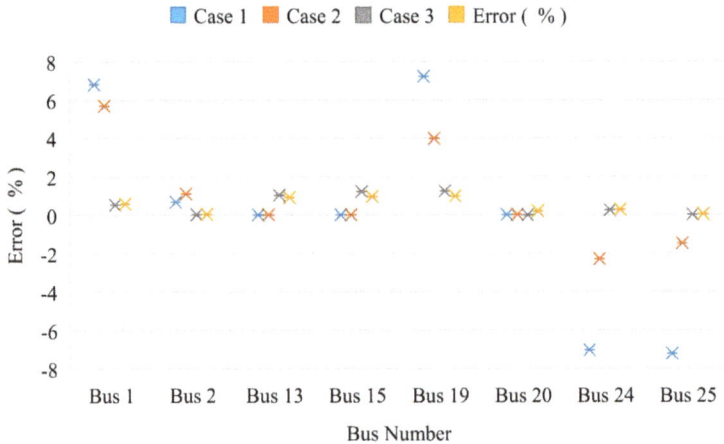

Figure 5. Comparison of errors in parameter estimation.

5.2. Worst Case Scenario Analysis

Under any operating conditions, all possible states have to be estimated to confirm the observability of a network. Therefore, the algorithm is tested in worst case scenarios, in which network states are likely to be violated. These scenarios are generally used during the planning phase of grid network operation, so it is known to DSOs [37,38]. Following worst case scenario (a), which is one of the worst operating scenarios, in which more DGs are integrated in the distribution network, other scenarios (b), (c) and (d) that could happen in real network operation have been set up to test the applicability of the method:

a.   Case 5 (C5): Minimum load and maximum generation. Minimum load (30% of maximum load at each bus) and maximum generation (20% of maximum load at each bus where generation is available) respectively.

b.   Case 6 (C6): Parameter estimation when line flow measurements (M4, M5 and M6) have an error of plus 5% of nominal reading.

c.   Case 7 (C7): Parameter estimation with large error on pseudo measurements models (over forecasting of load by 10%).

Network measurement setup is designated as a fully observable condition, i.e., as in case 4, which is then re-simulated to estimate the network parameters for the scenarios mentioned in cases 5–7; a snapshot of the results for voltage estimation is shown in Figures 6 and 7.

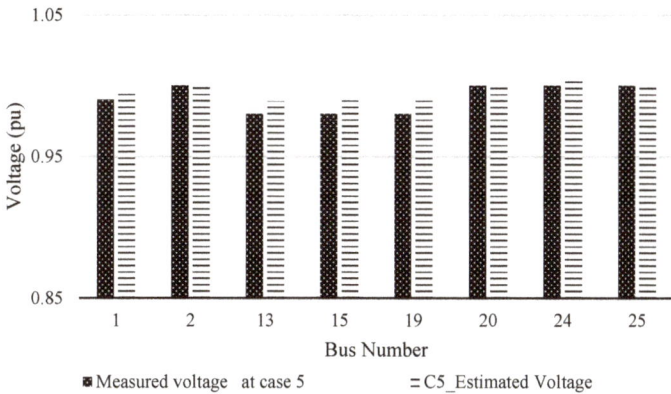

Figure 6. Comparison of estimated voltages for case 5.

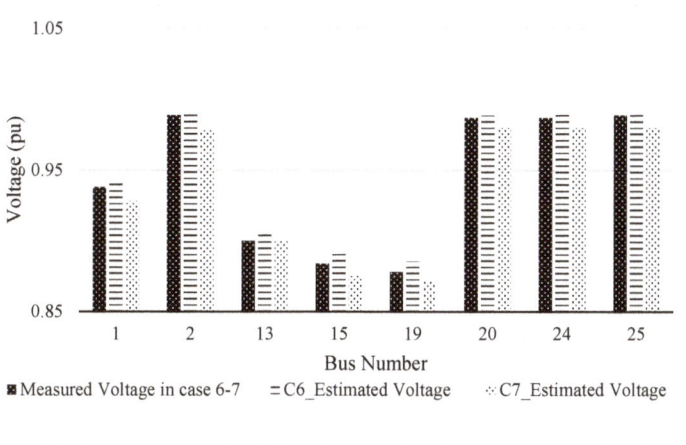

Figure 7. Comparison of estimated voltages for cases 6 and 7.

As seen in Figure 6, i.e., in case 5, it is found that using the same technique, network parameters can be estimated at close to their true value (maximum error 0.97%) even when the network is more active (i.e., in the presence of DG). About 0.01% more error than case 4 is noticed in case 5. This is because the estimator is modelled using net injection measurements at each bus, which logically include the impact of DG. However, a change in the system state due to RE penetration can result in the WLS estimator becoming trapped in local minima, and can add some error if the network is large, with the highest number of DGs [6]. From Figure 7, it can be seen that different patterns of impact are seen in case 6, i.e., only voltage estimates close to erroneous measurement points are more influenced (buses 13, 15, and 19). This is because the accuracy of the estimator is inversely proportional to the level of measurement error, and this will be reflected to the estimated states, which are close to the erroneous meter; furthermore, the effect could also be different with different types of measurements [39]. However, negligible impact is seen in the other buses, which are further away from the erroneous meters. The level of impact depends on the closeness of the meter and the number and locations of available meters in the same feeder [6,39]. This is valid for a radial feeder setup, which is the predominant case in most power distribution networks. The impact of forecasting error (case 7) can be reflected in the estimation error to some extent. This is because most of the pseudo measurements (load and generation at each bus) will be noisy and result in erroneous injection measurements in all buses. Therefore, estimated voltage magnitude will be lower than the real value in some buses due to the application of overestimated forecasting. As shown in Figure 7, since the error for the forecasted load is plus 10% and the estimating variable is the voltage, not the load, at each bus, the maximum error recorded on voltage estimation in this situation (case 7) is only 1.1%. The magnitude of error on the estimated parameter depends on the number of pseudomeasurements used by the estimator that have been selected from an incorrectly forecasted load/generation.

Key results are summarized in Table 2, showing the advantage of the proposed methodology over the conventional method. This shows the possibility of achieving higher accuracy in estimated states with the minimum use of real measurements (26%). The maximum error in the estimated states without applying the proposed technique is improved to 0.91 % (without considering DG) and 0.92% (considering DG). Even by using only 22% real measurements, all network states can be estimated with a maximum error of 1.14%. This proposed technique is a simple algebraic technique that identifies the minimum number of meters to be installed for full network observability. Therefore, it will be more economical and reliable than more complex methods, such as the Fisher information-based meter placement technique described in [9], which uses the pre-specified number of additional measurement units from the set of candidate units. The highest-quality result (minimum SE error) is obtained with 23% real measurements (20 real measuring units, with each unit consisting of 3 real measurements, i.e., with 60 real measurements out of 260 measurements (both real and pseudo)) in the test network. Even though in the latter case, about 23% real measurements are used, technique followed is comparatively more complex than the proposed method.

Table 2. Comparison of key results.

Comparative Study	Bus with Maximum SE Error	Measured Voltage (per unit)	Estimated Voltage (per unit)	Maximum SE Error Recorded in the Network (%)
Initial setup with only basic available measurements, i.e., case 1	25	0.989	0.917	7.28
Without applying proposed method (with 30% real measurements) i.e., case 2	1	0.938	0.991	5.65
Applying proposed method (with 22% real measurements), i.e., case 3	19	0.878	0.888	1.14
Using proposed method (with 26% real measurements) without considering DG penetration, i.e., case 4	19	0.878	0.886	0.91
Using proposed method (with 26% real measurements) with DG penetration, i.e., case 5	19	0.98	0.989	0.92

6. Conclusions and Future Works

This paper proposed an enhanced numerical state estimation assessment for distribution networks using a minimal number of measurements. The proposed method determines critical data sets to be measured and modelled as real and pseudo measurements and identifies the network observability status using a bus prioritization technique followed by the calculation of network states and the identification of its accuracy. This is the key contribution of this paper. It is possible to filter out unnecessary measurements and minimize the iteration cycle based on the bus priority function and added pre-assessment. The advantage of this method is the analysis of the trade-off between the number of real measurements for gaining observability and the accuracy of the estimated states. Thus,

Energies **2019**, *12*, 2230

based on the required level of accuracy and application of network states, a minimal number of real measurements can be selected. Various case studies were performed, demonstrating the applicability of the proposed method in network state estimation.

The proposed methodology can be useful for identifying the minimum number of any type of measurement devices for network observability, for example, PMU, RTDS, smart meters, etc. For instance, inclusion of PMU will not have any impact on the observability analysis proposed in this paper, since the algorithm is based on the number and locations of the measurements, not on their dynamic behavior. However, the proposed bus prioritization technique can be useful for identifying the critical locations at which to place the PMUs in the network, not only improving the accuracy of state estimation, but also defending against data injection attacks with minimum cost [21]. This could be possible either by treating PMU measurements as additional measurements in the traditional data set (this will add the computational burden) or the using distributed computation of mixed/hybrid state estimation [40]. Future work will focus on the integration of the observability function and state estimation module with control blocks for application in real-time network control in real grids using various measurement setups available in modern power systems.

Author Contributions: B.R.P. is the main author of this article and prepared the draft. All co-authors have provided technical input, reviewed the draft and supplied technical supervision for the work. They also carried out the project administration and funding acquisition for the project.

Funding: This work is funded by Energinet.dk, the Danish transmission system operator, under PSOForskEL programme, grant no. 12414, for the project DECODE.

Acknowledgments: The authors thankfully admit the contributions of ABB, Eniig, Syd Energi and Kenergy from Denmark as well as Smart Lab, University of Saskatchewan, Canada for their feedback and input during different project meetings and discussions.

Conflicts of Interest: The authors declare no conflict of interest.

References

1. Tornbjerg, J. Smart Distribution Grids Power Europe's Transition to Green Energy DSOs-the Backbone of the Energy Transition. Danish Energy Association. 2017. Available online: https://www.danskenergi.dk/sites/danskenergi.dk/files/media/dokumenter/2017-11/DSO_Magazine_210x297_ENG_V10.pdf (accessed on 27 May 2019).
2. Baran, M.E. Challenges in state estimation on distribution systems. In Proceedings of the IEEE Power Engineering Society Summer Meeting, Vancouver, BC, Canada, 15–19 July 2001.
3. Gelagaev, R.; Vermeyen, P.; Vandewalle, J.; Driesen, J. Numerical observability analysis of distribution systems. In Proceedings of the 14th International Conference on Harmonics and Quality of Power (ICHQP), Bergamo, Italy, 26–29 September 2010; pp. 1–6.
4. Andreassen, J. SCADA/DMS/AMI/NIS: Obsolete? 2017. Available online: http://blogs.esmartsystems.com/scada-dms-ami-nis-obsolete (accessed on 18 October 2017).
5. Abur, A.; Expósito, A.G. *Power System State Estimation: Theory and Implementation*; Marcel Dekker: New York, NY, USA, 2004.
6. Dehghanpour, K.; Wang, Z.; Wang, J.; Yuan, Y.; Bu, F. A survey on state estimation techniques and challenges in smart distribution systems. *IEEE Trans. Smart Grid* **2019**, *10*, 2312–2322. [CrossRef]
7. Bhela, S.; Kekatos, V.; Veeramachaneni, S. Enhancing observability in distribution grids using smart meter data. *IEEE Trans. Smart Grid* **2018**, *9*, 5953–5961. [CrossRef]
8. U.S. Energy Information Administration (EIA). Available online: https://www.eia.gov/tools/faqs/faq.php?id=108&t=3 (accessed on 8 October 2018).
9. Xygkis, T.C.; Korres, G.N.; Manousakis, N.M. Fisher information-based meter placement in distribution grids via the D-optimal experimental design. *IEEE Trans. Smart Grid* **2018**, *9*, 1452–1461. [CrossRef]
10. Singh, R.; Pal, B.C.; Vinter, R.B. Measurement Placement in Distribution System State Estimation. *IEEE Trans. Power Syst.* **2009**, *24*, 668–675. [CrossRef]
11. Heydt, G.T. The Next Generation of Power Distribution Systems. *IEEE Trans. Smart Grid* **2010**, *1*, 225–235. [CrossRef]

12. della Giustina, D.; Pau, M.; Pegoraro, P.; Ponci, F.; Sulis, S. Electrical distribution system state estimation: Measurement issues and challenges. *IEEE Instrum. Meas. Mag.* **2014**, *17*, 36–42. [CrossRef]

13. Huang, Y.-F.; Werner, S.; Huang, J.; Kashyap, N.; Gupta, V. State Estimation in Electric Power Grids: Meeting New Challenges Presented by the Requirements of the Future Grid. *IEEE Signal Process. Mag.* **2012**, *29*, 33–43. [CrossRef]

14. Liao, H.; Milanovic, J.V. Pathway to cost-efficient state estimation of future distribution networks. In Proceedings of the 2016 IEEE Power and Energy Society General Meeting (PESGM), Boston, MA, USA, 17–21 July 2016; pp. 1–5.

15. Primadianto, A.; Lu, C.N. A Review on Distribution System State Estimation. *IEEE Trans. Power Syst.* **2017**, *32*, 3875–3883. [CrossRef]

16. Pau, M.; Pegoraro, P.A.; Sulis, S. Efficient branch-current-based distribution system state estimation including synchronized measurements. *IEEE Trans. Instrum. Meas.* **2013**, *62*, 2419–2429. [CrossRef]

17. Pau, M.; Ponci, F.; Monti, A.; Sulis, S.; Muscas, C.; Pegoraro, P.A. An Efficient and Accurate Solution for Distribution System State Estimation with Multiarea Architecture. *IEEE Trans. Instrum. Meas.* **2017**, *66*, 910–919. [CrossRef]

18. Chakrabarti, S.; Kyriakides, E. PMU measurement uncertainty considerations in WLS state estimation. *IEEE Trans. Power Syst.* **2009**, *24*, 1062–1071. [CrossRef]

19. Asprou, M.; Kyriakides, E. The effect of PMU measurement delays on a linear state estimator: Centralized Vs decentralized. In Proceedings of the 2017 IEEE Power & Energy Society General Meeting, Chicago, IL, USA, 16–20 July 2017; pp. 1–5.

20. Thirugnanasambandam, V.; Jain, T. Placement of Synchronized Measurements for Power System Observability during Cascaded Outages. *Int. J. Emerg. Electr. Power Syst.* **2017**, *18*, 12–19. [CrossRef]

21. Yang, Q.; Jiang, L.; Hao, W.; Zhou, B.; Yang, P.; Lv, Z. PMU Placement in Electric Transmission Networks for Reliable State Estimation Against False Data Injection Attacks. *IEEE Internet Things J.* **2017**, *4*, 1978–1986. [CrossRef]

22. Chakrabarti, S.; Kyriakides, E.; Ledwich, G.; Ghosh, A. Inclusion of PMU current phasor measurements in a power system state estimator. *IET Gener. Transm. Distrib.* **2010**, *4*, 1104. [CrossRef]

23. Pegoraro, P.A.; Sulis, S. Robustness-Oriented Meter Placement for Distribution System State Estimation in Presence of Network Parameter Uncertainty. *IEEE Trans. Instrum. Meas.* **2013**, *62*, 954–962. [CrossRef]

24. Liu, J.; Ponci, F.; Monti, A.; Muscas, C.; Pegoraro, P.A.; Sulis, S. Optimal Meter Placement for Robust Measurement Systems in Active Distribution Grids. *IEEE Trans. Instrum. Meas.* **2014**, *63*, 1096–1105. [CrossRef]

25. Pegoraro, P.A.; Meloni, A.; Atzori, L.; Castello, P.; Sulis, S. PMU-Based Distribution System State Estimation with Adaptive Accuracy Exploiting Local Decision Metrics and IoT Paradigm. *IEEE Trans. Instrum. Meas.* **2017**, *66*, 704–714. [CrossRef]

26. Muscas, C.; Pau, M.; Pegoraro, P.A.; Sulis, S. Effects of Measurements and Pseudomeasurements Correlation in Distribution System State Estimation. *IEEE Trans. Instrum. Meas.* **2014**, *63*, 2813–2823. [CrossRef]

27. Celli, G.; Pegoraro, P.A.; Pilo, F.; Pisano, G.; Sulis, S. DMS Cyber-Physical Simulation for Assessing the Impact of State Estimation and Communication Media in Smart Grid Operation. *IEEE Trans. Power Syst.* **2014**, *29*, 2436–2446. [CrossRef]

28. Pokhrel, B.R.; Karthikeyan, N.; Bak-Jensen, B.; Pillai, J.R. Effect of Smart Meter Measurements Data On Distribution State Estimation. In Proceedings of the 19th IEEE International Conference on Industrial Technology (ICIT), Lyon, France, 20–22 February 2018; pp. 1207–1212.

29. Pokhrel, B.R.; Karthikeyan, N.; Bak-Jensen, B.; Pillai, J.R.; Mazhari, S.M.; Chung, C.Y. An Intelligent Approach to Observability of Distribution Networks. In Proceedings of the IEEE PES General Meeting, Portland, OR, USA, 5–10 August 2018.

30. Stevenson, W.D. *Elements of Power System Analysis*; McGraw-Hill: New York, NY, USA, 1982.

31. Brinkmann, B.; Member, S.; Negnevitsky, M.; Member, S. A Probabilistic Approach to Observability of Distribution Networks. *IEEE Trans. Power Syst.* **2017**, *32*, 1169–1178. [CrossRef]

32. Methods and Formulas for Probability Density Function. Available online: https://support.minitab.com/en-us/minitab-express/1/help-and-how-to/basic-statistics/probability-distributions/how-to/probability-density-function-pdf/methods-and-formulas/methods-and-formulas/ (accessed on 16 November 2018).

33. Karthikeyan, N.; Pokhrel, B.R.; Pillai, J.R.; Bak-jensen, B.; Jensen, A.; Helth, T.; Andreasen, J.; Frederiksen, K.H.B. Coordinated Control with Improved Observability for Network Congestion Management in Medium-Voltage Distribution Grid. In Proceedings of the CIGRE Session, Paris, France, 26–31 August 2018.

34. Quan, H.; Srinivasan, D.; Khosravi, A. Short-Term Load and Wind Power Forecasting Using Neural Network-Based Prediction Intervals. *IEEE Trans. Neural Networks Learn. Syst.* **2014**, *25*, 303–315. [CrossRef] [PubMed]

35. Regulations for Grid Connection|Energinet. Available online: https://en.energinet.dk/Electricity/Rules-and-Regulations/Regulations-for-grid-connection (accessed on 16 November 2018).

36. IEC 62053-11. Available online: https://webstore.iec.ch/publication/26235 (accessed on 6 November 2018).

37. Kulmala, A.; Repo, S.; Jarventausta, P. Using statistical distribution network planning for voltage control method selection. In Proceedings of the IET Conference on Renewable Power Generation (RPG 2011), Edinburgh, UK, 6–8 September 2011; pp. 41–41.

38. Celli, G.; Pilo, F.; Soma, G.G.; Cicoria, R.; Mauri, G.; Fasciolo, E.; Fogliata, G. A comparison of distribution network planning solutions: Traditional reinforcement versus integration of distributed energy storage. In Proceedings of the 2013 IEEE Grenoble Conference, Grenoble, France, 16–20 June 2013; pp. 1–6.

39. Sabri, N.C.O.; Majdoub, M.; Cheddadi, B.; Belfqih, A.; Boukherouaa, J.; El Mariami, F. Correlation between Weight, Error and Type of Measurements in WLS State Estimation of a Real Network. Available online: https://www.researchgate.net/publication/318582394_Correlation_between_weight_error_and_type_of_measurements_in_WLS_state_estimation_of_a_real_network (accessed on 28 May 2019).

40. Sangrody, H.A.; Ameli, M.T.; Meshkatoddini, M.R. The effect of phasor measurement units on the accuracy of the network estimated variables. In Proceedings of the 2009 Second International Conference on Developments in eSystems Engineering, Abu Dhabi, UAE, 14–16 December 2009; pp. 66–71.

MDPI

St. Alban-Anlage 66

4052 Basel

Switzerland

Tel. +41 61 683 77 34

Fax +41 61 302 89 18

www.mdpi.com

Energies Editorial Office

E-mail: energies@mdpi.com

www.mdpi.com/journal/energies

www.ingramcontent.com/pod-product-compliance
Lightning Source LLC
Chambersburg PA
CBHW051914210326

41597CB00033B/6136